AN INTRODUCTION TO PROOF AND ANALYSIS

Exploring the INFINITE

TEXTBOOKS in MATHEMATICS

Series Editors: Al Boggess and Ken Rosen

PUBLISHED TITLES

ABSTRACT ALGEBRA: A GENTLE INTRODUCTION
Gary L. Mullen and James A. Sellers

ABSTRACT ALGEBRA: AN INTERACTIVE APPROACH, SECOND EDITION
William Paulsen

ABSTRACT ALGEBRA: AN INQUIRY-BASED APPROACH
Jonathan K. Hodge, Steven Schlicker, and Ted Sundstrom

ADVANCED LINEAR ALGEBRA
Hugo Woerdeman

APPLIED ABSTRACT ALGEBRA WITH MAPLE™ AND MATLAB®, THIRD EDITION
Richard Klima, Neil Sigmon, and Ernest Stitzinger

APPLIED DIFFERENTIAL EQUATIONS: THE PRIMARY COURSE
Vladimir Dobrushkin

A BRIDGE TO HIGHER MATHEMATICS
Valentin Deaconu and Donald C. Pfaff

COMPUTATIONAL MATHEMATICS: MODELS, METHODS, AND ANALYSIS WITH MATLAB® AND MPI, SECOND EDITION
Robert E. White

DIFFERENTIAL EQUATIONS: THEORY, TECHNIQUE, AND PRACTICE, SECOND EDITION
Steven G. Krantz

DIFFERENTIAL EQUATIONS: THEORY, TECHNIQUE, AND PRACTICE WITH BOUNDARY VALUE PROBLEMS
Steven G. Krantz

DIFFERENTIAL EQUATIONS WITH APPLICATIONS AND HISTORICAL NOTES, THIRD EDITION
George F. Simmons

DIFFERENTIAL EQUATIONS WITH MATLAB®: EXPLORATION, APPLICATIONS, AND THEORY
Mark A. McKibben and Micah D. Webster

ELEMENTARY NUMBER THEORY
James S. Kraft and Lawrence C. Washington

PUBLISHED TITLES CONTINUED

EXPLORING CALCULUS: LABS AND PROJECTS WITH MATHEMATICA®
Crista Arangala and Karen A. Yokley

EXPLORING GEOMETRY, SECOND EDITION
Michael Hvidsten

EXPLORING LINEAR ALGEBRA: LABS AND PROJECTS WITH MATHEMATICA®
Crista Arangala

GRAPHS & DIGRAPHS, SIXTH EDITION
Gary Chartrand, Linda Lesniak, and Ping Zhang

INTRODUCTION TO ABSTRACT ALGEBRA, SECOND EDITION
Jonathan D. H. Smith

INTRODUCTION TO MATHEMATICAL PROOFS: A TRANSITION TO ADVANCED MATHEMATICS, SECOND EDITION
Charles E. Roberts, Jr.

INTRODUCTION TO NUMBER THEORY, SECOND EDITION
Marty Erickson, Anthony Vazzana, and David Garth

LINEAR ALGEBRA, GEOMETRY AND TRANSFORMATION
Bruce Solomon

MATHEMATICAL MODELLING WITH CASE STUDIES: USING MAPLE™ AND MATLAB®, THIRD EDITION
B. Barnes and G. R. Fulford

MATHEMATICS IN GAMES, SPORTS, AND GAMBLING–THE GAMES PEOPLE PLAY, SECOND EDITION
Ronald J. Gould

THE MATHEMATICS OF GAMES: AN INTRODUCTION TO PROBABILITY
David G. Taylor

A MATLAB® COMPANION TO COMPLEX VARIABLES
A. David Wunsch

MEASURE THEORY AND FINE PROPERTIES OF FUNCTIONS, REVISED EDITION
Lawrence C. Evans and Ronald F. Gariepy

NUMERICAL ANALYSIS FOR ENGINEERS: METHODS AND APPLICATIONS, SECOND EDITION
Bilal Ayyub and Richard H. McCuen

ORDINARY DIFFERENTIAL EQUATIONS: AN INTRODUCTION TO THE FUNDAMENTALS
Kenneth B. Howell

PRINCIPLES OF FOURIER ANALYSIS, SECOND EDITION
Kenneth B. Howell

PUBLISHED TITLES CONTINUED

RISK ANALYSIS IN ENGINEERING AND ECONOMICS, SECOND EDITION
Bilal M. Ayyub

SPORTS MATH: AN INTRODUCTORY COURSE IN THE MATHEMATICS OF SPORTS SCIENCE AND SPORTS ANALYTICS
Roland B. Minton

TRANSFORMATIONAL PLANE GEOMETRY
Ronald N. Umble and Zhigang Han

TEXTBOOKS in MATHEMATICS

AN INTRODUCTION TO PROOF AND ANALYSIS

Exploring the INFINITE

Jennifer Brooks
University of Montana, USA

CRC Press is an imprint of the
Taylor & Francis Group an **informa** business
A CHAPMAN & HALL BOOK

CRC Press
Taylor & Francis Group
6000 Broken Sound Parkway NW, Suite 300
Boca Raton, FL 33487-2742

First issued in paperback 2022

Version Date: 20161019

ISBN 13: 978-1-03-247704-6 (pbk)
ISBN 13: 978-1-4987-0449-6 (hbk)

DOI: 10.1201/9781315371719

Publisher's Note
The publisher has gone to great lengths to ensure the quality of this reprint but points out that some imperfections in the original copies may be apparent.

Visit the Taylor & Francis Web site at
http://www.taylorandfrancis.com

and the CRC Press Web site at
http://www.crcpress.com

Contents

Preface

Mathematics is an incredibly broad endeavor, encompassing areas as diverse as number theory, numerical analysis, graph theory, and topology, just to name a few. What holds the discipline together as a single, recognizable unit are its quest to identify and abstract patterns and its reliance on logical argumentation.

This book guides the reader through these processes of abstraction and logical argumentation, to make the transition from student of mathematics to practitioner of mathematics. Such a transition requires more than knowledge of the definitions of mathematical structures, elementary logic, and standard proof techniques. Indeed, the student focused on only these will develop little more than the ability (common among young students but nearly useless in research) to identify a number of proof templates and to apply them in predictable ways to standard problems. This book aims to do something more; it aims to help readers learn to explore mathematical situations, to make conjectures, and only then to apply methods of proof. Practitioners of mathematics must do all of these things.

How do we teach exploration? First, we provide many opportunities for creative mathematical thought. Whenever possible, we pose questions for readers to explore, hopefully ending in a proof of a conjecture, rather than suggesting the result to prove. Thus we would ask readers to find and prove a formula for the sum of the first n natural numbers rather than asking for a proof that $1+2+3+\ldots+n = n(n+1)/2$. Second, we help readers develop an invaluable skill for mathematical exploration, namely basic facility with programming.

Learning to program does more than help with the exploration of a mathematical situation; it may also lead from the conjecture phase toward a valid proof. For example, suppose we are given a continuous function $f\colon [0,1] \to \mathbb{R}$ such that $f(0) < 0$ and $f(1) > 0$ and we want an x between 0 and 1 for which $f(x) = 0$. We could hit on the idea of finding it with a simple divide-and-conquer program: start with $a_0 = 0$ and $b_0 = 1$. $f(a_0) < 0$ and $f(b_0) > 0$ by hypothesis. Now consider their average, $c_0 = \frac{1}{2}$. If $f(c_0) = 0$, we have found our x. Suppose we aren't so lucky. If $f(c_0) < 0$, set $a_1 = \frac{1}{2}$ and $b_1 = 1$. Otherwise set $a_1 = 0$ and $b_1 = \frac{1}{2}$. Now repeat the process; $f(a_1) < 0$ and $f(b_1) > 0$. Consider their average $c_1 = \frac{a_1+b_1}{2}$. Based on the value $f(c_1)$, define a_2 and b_2 so that f still changes sign on the new, smaller interval $[a_2, b_2]$. Writing a computer program to carry out this algorithm is trivial. *But we have also found a viable approach to a general proof of the Intermediate Value Theorem.*

Prerequisites; Ground rules An awkward feature of all books of this type is that one must decide what readers are expected to know and what they can assume as known in their proofs. Our approach is to avoid pedantry, perhaps at the cost of some rigor. We believe that proof is better presented as "a clear, logical explanation of the solution to an interesting non-obvious question" than as "the construction of a formal argument using only axioms, definitions, and previously-established theorems." It is not that we believe that the latter is incorrect, but rather that we have seen students become almost paralyzed in their early proof-writing efforts, not sure if they are allowed to assume that, say, the basic rules of algebra are valid. We would rather readers think more about the mathematical process of exploring, conjecturing, and proving, and less about the formal rules.

That said, here are our ground rules: we assume readers know that the real number system satisfies the ordered field axioms (even if they don't recognize the term "ordered field"). Thus the basic rules learned in school for doing arithmetic and for solving equations and inequalities are indeed valid. We will briefly review the properties of familiar number systems in Chapter 1, but after Chapter 1 we will use them without comment. Later in the book, we will step back and look more closely at how one can build up the field of rational numbers from the integers and how one can construct the real numbers from the rational numbers. At that time, we may pause to consider again why all of these rules work.

Format of the book The chapters of this text are divided into two parts. The first part serves as an introduction to proof and abstract mathematics and aims to prepare the reader for advanced course work in all areas of mathematics. It thus includes all the standard material from a "transition to proof" course: induction, basic logic, elementary set theory, functions and relations, and basic number theory and combinatorics. Part II constitutes an introduction to the basic concepts of analysis, including limits of sequences of real numbers and of functions, infinite series, the structure of the real line, and continuous functions.

That said, we aim for a fresh perspective; we achieve additional coherence by focusing on exploring the infinite, as the title suggests. Thus in Part I, we emphasize sequences, recurrence relations, and the cardinality of infinite sets as our main applications of induction, sets, and functions. In Part II, the infinite is everywhere, as a first analysis course is primarily about limiting processes. Still, we emphasize sequences. We think of real numbers as those objects that can be approximated by Cauchy sequences of rational numbers. We define continuous functions as those that send convergent sequences to convergent sequences. And we define compact sets as those with the property that every sequence in the set has a subsequence converging to a point of the set. We leave more abstract topological notions for a future course. This approach to analysis is somewhat non-standard; often a first analysis course is presented as "single-variable calculus made rigorous." Such a course would certainly cover not only limits and continuity, but also differentiation and integration. Such material is an important part of the undergraduate curriculum. We choose, however, to worry less about whether we cover all of calculus once again and focus instead on understanding the idea of a limit at a very deep level. It has been our observation from teaching this material that students understand limits more easily in the context of sequences (as opposed to functions), and thus time spent here is time well spent. An additional advantage of our focus on sequences is that it allows readers to continue to use their new programming skills to explore concepts in analysis. Many theorems in analysis can be proved by inductively constructing sequences whose limit is an object of interest (the supremum of a set, the zero of a continuous function, the fixed point of a contractive mapping, etc.). If one can write a program to execute such an algorithm, one has come a long way towards understanding the result and its proof.

How to use this book This book is designed not only to be readable, but to help students learn how to read math books. Each section is a combination of exposition, motivation, examples, and embedded exercises. Readers should do all these exercises, as they ask them to use new definitions, practice new techniques, fill in details of a calculation, or prove corollaries and extensions of results in the text. The point is that reading mathematics is a very active endeavor, not at all like reading other books. It is a slow process, requiring paper, pencil, and active struggle.

The text also includes many problems at the ends of chapters. Unlike the exercises, which are focused and explicit, the problems encourage exploration. In a problem, we pose a question but seldom say what the result is; the reader must first find the result and then prove it. Such problems do have more potential to frustrate than requests for proofs of specific statements. But doing mathematics is at least as much about trying to figure out what is true as it is about trying to find a good proof of a known result.

Acknowledgements First, I would like to thank Jeffrey Kacmarcik. He is a Senior Systems Programmer at the University of Wisconsin and consulted with me on the Appendix and the Python code throughout the book. I thank John D'Angelo for his encouragement, advice, and feedback. I thank my daughter, the joy of my life, for her love and friendship. Finally, I thank Curtis for his love and support, and for his companionship on this wonderful path we have chosen.

Part I

Fundamentals of Abstract Mathematics

Chapter 1

Basic Notions

If you are reading this chapter, you are probably beginning a course with a title like *Introduction to Abstract Mathematics* or *Mathematical Reasoning*. Such courses are designed to help you make a transition from courses like calculus that focus mainly on developing the computational fluency needed by a *user* of mathematics to advanced courses that focus on developing the understanding of the theoretical underpinnings of mathematics needed by a *creator* of mathematics. Where should such a course begin? If it is an "introductory" course, perhaps it should start at the beginning, not assuming any particular mathematical knowledge. In a sense, we do this; we do not assume you have any experience with reading or writing mathematical proofs. On the other hand, this is far from your first course in mathematics. At this point, you are certainly fluent in arithmetic and algebra and very comfortable with trigonometry, calculus, and basic plane geometry. This course would feel like some formal game if we asked you to pretend you do not have all this knowledge.

Here is what we do ask: think of your journey through mathematics as a journey up a spiral staircase. In a sense, each time around you seem to be moving over the same territory, but in fact each time you come upon it, you do so at a higher level. In this chapter, we discuss what facts about number systems you can freely use as you embark on this stage of the journey. Later in the book, however, we may circle back around and ask you to think more deeply about these basic notions and justify statements that you had previously taken as facts.

1.1 A First Look at Some Familiar Number Systems

1.1.1 Integers and natural numbers

We let $\mathbb{Z}$ denote the set of integers $\{\ldots, -3, -2, -1, 0, 1, 2, 3, \ldots\}$. This set is *closed* under the operations of addition $(+)$ and multiplication $(\cdot)$. In other words, when we add two integers, the result is a unique integer, and when we multiply two integers, the result is a unique integer.

We must know what properties these operations satisfy. Even if you have never seen a formal enumeration of these properties, you have freely used them for many years. The first two properties, for example, imply that if you want to add up a list of numbers, you can add them in any order you like and group them in any way you like. Here's the complete list.

1. The operations of addition and multiplication of integers are *associative.* That is, if m, n, p are integers, $(m+n)+p = m+(n+p)$ and $(m \cdot n) \cdot p = m \cdot (n \cdot p)$.

2. The operations of addition and multiplication of integers are *commutative.* That is, if m, n are integers, $m + n = n + m$ and $m \cdot n = n \cdot m$.

3. There is a unique integer, 0, such that $n+0 = 0+n = n$ for all integers n. Furthermore, there is a unique integer, 1, such that $n \cdot 1 = 1 \cdot n = n$ for all integers n. 0 is called the *additive identity* and 1 is called the *multiplicative identity.*

4. Every integer has an *additive inverse.* That is, if n is an integer, there is a unique integer m such that $n + m = m + n = 0$. We write $-n$ for the unique additive inverse of n.

5. Multiplication *distributes* over addition. That is, if m, n, p are integers, $m \cdot (n+p) = m \cdot n + m \cdot p$ and $(m+n) \cdot p = m \cdot p + n \cdot p$.

When no confusion can arise, we will use juxtaposition to denote multiplication of integers, omitting the dot. Thus the distributive law can also be written

$$m(n+p) = mn + mp.$$

We will also freely use $n - m$ to mean $n + (-m)$, and we will use exponents to indicate repeated multiplication (for example, n^2 for $n \cdot n$, etc.).

We may use this list to obtain other *propositions* about the integers. For example, consider the following proposition:

Proposition 1.1. *For all integers n, $0 \cdot n = 0$.*

Of course, we all consider the above proposition to be true. In this book, however, we don't simply want to appeal to past experience or authority. Instead, we want to demonstrate that a given statement follows from other statements we have already agreed are true. Such a demonstration is a *proof.* Let's give the simple proof using the properties of integer operations given above.

Proof. Property 3 states that the integers contain a *unique* additive identity, 0. Thus to show that $0 \cdot n = 0$, it suffices to show that $0 \cdot n$ has the same property characterizing the additive identity. Observe that, for any integer n,

$$\begin{aligned} n + 0 \cdot n &= 1 \cdot n + 0 \cdot n && \text{(by Property 3)} \\ &= (1+0) \cdot n && \text{(by Property 5)} \\ &= 1 \cdot n && \text{(by Property 3)} \\ &= n && \text{(by Property 3).} \end{aligned}$$

Thus $0 \cdot n = 0$, as claimed. □

To summarize, in order to establish a new proposition about operations with integers, we must give a proof using only the Properties 1–5 and previously-proved propositions.

Exercise 1.1. Prove each of the following propositions about integers using only Properties 1–5 above and Proposition 1.1.

(a) For all integers m and n, $(m+n)^2 = m^2 + 2mn + n^2$.

(b) For all integers n, $(-1) \cdot n = -n$. Before proving this statement, explain carefully what $(-1) \cdot n$ means and what $-n$ means.

(c) For all integers m and n, $(-m)n = -(mn)$.

(d) For all integers n, $(-n)^2 = n^2$.

(e) For all integers m and n, $(m-n)^2 = m^2 - 2mn + n^2$.

What do we do if we are confronted with a proposition about operations with integers that we suspect is false? In a sense, this situation is easier; we need only supply one concrete counterexample. Suppose, for example, a classmate claims that, for all integers m and n,

$$(m+n)^4 = m^4 + m^3 n + m^2 n^2 + mn^3 + n^4.$$

All we need to do to disprove the claim is to find integers m and n for which equality fails. For instance, if we take $m = n = 1$, the left-hand side equals 16 whereas the right-hand side equals 5.

Natural numbers. The set $\mathbb{N} = \{1, 2, 3, \ldots\}$ of *natural numbers* gets special attention. The natural numbers are convenient for counting and for labeling terms in sequences and elements in sets. $\mathbb{N}$ is closed under addition and multiplication, but several of the other algebraic properties of the integers are not satisfied; $\mathbb{N}$ does not contain an additive identity, and elements of $\mathbb{N}$ do not have additive inverses in $\mathbb{N}$.

Divisibility and the Fundamental Theorem of Arithmetic. To this point, in our discussion of operations with integers, we have talked about addition, subtraction, and multiplication, but not division. The reason is that, unlike the other operations, which take a pair of integers and return another integer, dividing one integer by another does not always return an integer. Division is, nonetheless, an important operation.

Definition 1.1. Let m and n be integers, with $m \neq 0$. We say that m **divides** n (or that n **is divisible by** m) if there exists an integer k such that $n = mk$. When m divides n, we call m and k **factors** or **divisors** of n.

In the special case in which n is divisible by 2, we say that n is *even*. In this case, there exists an integer k such that $n = 2k$. If n is not even, it is *odd*, and there exists an integer k such that $n = 2k + 1$.

Exercise 1.2. Is the set of even integers closed under addition and multiplication? If so, give proofs. If not, provide counterexamples.

Exercise 1.3. Is the set of odd integers closed under addition and multiplication? If so, give proofs. If not, provide counterexamples.

When discussing divisibility, *prime* numbers play a particularly important role. Let's recall the definition.

Definition 1.2. A natural number $p > 1$ is **prime** if the only natural numbers dividing it are 1 and p.

Thus 7 is prime because its only positive divisors are 1 and 7, but 21 is not prime because 3 and 7 divide 21. For the moment, we accept the following important (dare we say, fundamental) theorem as fact.

Theorem 1.1 (Fundamental Theorem of Arithmetic). *Let n be a natural number. Then there exist primes $p_1, \ldots, p_d$, unique up to reordering, and unique natural numbers $a_1, \ldots, a_d$ such that*

$$n = p_1^{a_1} \cdots p_d^{a_d}.$$

The above is called the **prime factorization** *of n.*

Perhaps we should take a moment to talk about reading mathematics. Often when you read the statement of a theorem for the first time, you will find it hard to understand. Such statements often need to be read many times. A good way to try to understand what a theorem says is to figure out what it says for some concrete examples. Look again at the statement of the Fundamental Theorem of Arithmetic; it is a theorem about natural numbers. What does the theorem say, for example, about the natural number 300? Stripping away all the notation for a moment, it says simply that we can write our natural number as a product of powers of primes. For 300, we can write

$$300 = 3 \cdot 2^2 \cdot 5^2.$$

For a general natural number n, we can't say how many primes we will need. Thus in the statement of the theorem it just says there exist primes $p_1, \ldots, p_d$. For $n = 300$, we had $d = 3$. The theorem also says that, up to reordering, the primes are unique. So we could write $300 = 2^2 \cdot 3 \cdot 5^2$, but there is no way to write 300 without having two factors of 2, two factors of 5, and one factor of 3.

Statements of theorems are often very dense. You can come to understand them better if you read them several times, try to paraphrase them using simpler language, and work out what they say in the context of familiar examples.

1.1.2 Rational numbers and real numbers

Although every integer has an additive inverse in the integers, most integers do not have a multiplicative inverse in the integers. We thus form the set $\mathbb{Q}$ of *rational numbers.* Each x in $\mathbb{Q}$ can be written as a quotient of integers in which the denominator is not zero. As you are no doubt aware, however, the representation of a rational number as a ratio of integers is not unique. For example, $\frac{1}{2}$, $\frac{5}{10}$, and $\frac{-4}{-8}$ all represent the same rational number.

You are, of course, very familiar with addition and multiplication of rational numbers. The set $\mathbb{Q}$ together with these operations forms a *field.* The defining properties of a field are the same as the defining properties of the integers described above except that, in a field, every non-zero element has a *multiplicative inverse.* We summarize in the following definition:

Definition 1.3. A set F that is closed under two operations (addition and multiplication) is a **field** if the following properties are satisfied:

(F1) (Associativity of operations) For all x, y, z in F, $(x + y) + z = x + (y + z)$ and $(xy)z = x(yz)$.

(F2) (Commutativity of operations) For all x, y in F, $x+y = y+x$ and $xy = yx$.

(F3) (Existence of identities) There is a unique element of F, denoted 0, such that $x + 0 = 0 + x = x$ for all x in F. Furthermore, there is a unique element of F different from 0, denoted 1, such that $x \cdot 1 = 1 \cdot x = x$ for all x in F. 0 is the **additive identity** and 1 is the **multiplicative identity**.

(F4) (Existence of inverses) If x is in F, there exists a unique element of F, denoted $-x$, such that $x + (-x) = (-x) + x = 0$. Furthermore, if x is in F and $x \neq 0$, there exists a unique element of F, denoted x^{-1}, such that $xx^{-1} = x^{-1}x = 1$. $-x$ is the **additive inverse** of x and x^{-1} is the **multiplicative inverse** of x.

(F5) (Distributive property) For all x, y, z in F, $x(y+z) = xy+xz$ and $(x+y)z = xz + yz$.

We use the notation $x - y$ to mean $x + (-y)$ and we use $\frac{x}{y}$ to mean xy^{-1}. Because this list includes all the properties satisfied by operations with integers, all of the results above (Proposition 1.1 and Exercise 1.1) hold in any field as well.

The *real number system* $\mathbb{R}$ is another familiar number system. Defining $\mathbb{R}$ precisely requires considerable care; we defer to Chapter 9. You can think of it as those numbers with finite or infinite decimal representations. For now, we take the real number system as essentially known from calculus. The real numbers form a field that contains the field of rational numbers. Essentially, $\mathbb{R}$ includes the rational numbers together with any number that can be approximated to arbitrary accuracy by rational numbers. The familiar numbers π, e, and $\sqrt{2}$ are examples of real numbers that are not rational; we compute with them by using a decimal approximation with sufficient accuracy for our intended application. The fact that $\mathbb{R}$ contains all numbers that can be approximated by rational numbers and not just the rational numbers means that $\mathbb{R}$ has no "holes." This property of $\mathbb{R}$ is called *completeness*. As a consequence of completeness, $\mathbb{R}$ is the correct number system to use when we are dealing with quantities that we think of as varying along a continuum. It is thus the number system best suited for calculus.

Properties (F1)–(F5) above justify all the algebraic manipulations we perform when we simplify or expand expressions involving rational or real numbers. They also justify all the algebraic manipulations we do when we solve equations.

What do we do when we solve equations? The basic idea, as you well know, is to convert the equation we are trying to solve into an equivalent equation by adding the same thing to both sides or by multiplying both sides by the same non-zero number. The next proposition justifies this approach.

Proposition 1.2. *Let a, b, and c be real numbers.*

(a) $a = b$ if and only if $a + c = b + c$. (Thus we don't change the solutions to an equation if we add the same thing to both sides.)

(b) Suppose $c \neq 0$. $a = b$ if and only if $ca = cb$. (Thus we don't change the solutions to an equation if we multiply both sides by a non-zero quantity.)

Proof. We prove part (a) and leave the proof of part (b) to the reader.

The statement to prove is an *if and only if*, or *biconditional*, statement, and so we must actually prove two things. The first is easy; suppose that $a = b$. Then by simple substitution, $a + c = b + c$.

For the second statement, suppose $a + c = b + c$. Then

$$\begin{aligned}
a &= a + 0 && \text{(by F3)} \\
&= a + (c + (-c)) && \text{(by F4)} \\
&= (a + c) + (-c) && \text{(by F1)} \\
&= (b + c) + (-c) && \text{(by substitution)} \\
&= b + (c + (-c)) && \text{(by F1)} \\
&= b + 0 && \text{(by F4)} \\
&= b && \text{(by F3)},
\end{aligned}$$

as claimed. □

Exercise 1.4. Prove part (b) of Proposition 1.2.

We illustrate the use of Proposition 1.2 to solve a simple linear equation.

Example 1.1. Suppose we want to find a real number x satisfying $3x + 3 = 4$. Let's do the algebra, observing at each step which of the properties (F1)–(F5) we are using.

We begin by adding -3 to both sides of the equation because -3 is the additive inverse of 3. This step is legitimate by part (a) of Proposition 1.2. Of course, when we add -3 to the right-hand side, we do the arithmetic and get 1. When we add -3 to the left-hand side we get

$$\begin{aligned}(3x+3)+(-3) &= 3x+(3+(-3)) && \text{(by F1)}\\ &= 3x+0 && \text{(by F4)}\\ &= 3x && \text{(by F3).}\end{aligned}$$

Our original equation is thus equivalent to the equation $3x = 1$. Next, we want to multiply both sides of the equation by the multiplicative inverse of 3, which we denote by $\frac{1}{3}$. This step is legitimate by part (b) of Proposition 1.2. The left-hand side becomes

$$\begin{aligned}\frac{1}{3}(3x) &= \left(\frac{1}{3}\cdot 3\right)x && \text{(by F1)}\\ &= 1x && \text{(by F4)}\\ &= x && \text{(by F3).}\end{aligned}$$

When we multiply the right-hand side of the equation by $\frac{1}{3}$, we get

$$\frac{1}{3}\cdot 1 = \frac{1}{3},$$

by (F4). Thus the original equation is equivalent to $x = \frac{1}{3}$.

Exercise 1.5. Let x and y be elements of $\mathbb{R}$. Show that if $xy = 0$, then either $x = 0$ or $y = 0$. Here is how you might proceed: Suppose that $xy = 0$ but $x \neq 0$. Show that y must be 0.

Throughout the remaining chapters of this book, you should feel free to do algebra *without* explicitly justifying every step by referring to Properties (F1)–(F5).

1.2 Inequalities

To this point, we have studied arithmetic in $\mathbb{Z}$, $\mathbb{Q}$, and $\mathbb{R}$. In addition to these two operations, we typically define an *order relation* on each set, denoted $<$. For definiteness, let's think of the order relation as defined on $\mathbb{R}$. We then get an order relation on $\mathbb{Z}$ or on $\mathbb{Q}$ by restricting the order on $\mathbb{R}$ to these subsets.

To be an order relation on $\mathbb{R}$, $<$ has to satisfy three properties:

(ORD1) (Comparability) For all x, y in $\mathbb{R}$ with $x \neq y$, either $x < y$ or $y < x$.

(ORD2) (Non-reflexivity) There is no x for which $x < x$.

(ORD3) (Transitivity) For all x, y, z in $\mathbb{R}$, if $x < y$ and $y < z$, then $x < z$.

As is customary, we sometimes write $y > x$ instead of $x < y$, and we use $x \leq y$ (resp., $x \geq y$) to indicate that either $x < y$ (resp., $x > y$) or $x = y$.

The real numbers are, however, more than just a field with an order relation; $\mathbb{R}$ with the familiar order relation is actually an *ordered field.* In an ordered field, the order relation interacts in a certain way with the arithmetic operations.

Definition 1.4. Let F be a field with an order relation $<$. Let P be the set of all x in F satisfying $0 < x$. We call P the set of **positive elements** of F. F with the order relation $<$ is an **ordered field** if the following properties hold:

(OF1) (Trichotomy) For all x in F, precisely one of the following holds:

(a) x is in P

(b) $-x$ is in P

(c) $x = 0$.

(OF2) (Closure under addition) If x and y are in P, then $x + y$ is in P.

(OF3) (Closure under multiplication) If x and y are in P, then xy is in P.

As an illustration of the order axioms, we prove that the square of any real number is non-negative.

Proposition 1.3. *Let F be an ordered field. If x is in F, then either $x^2 = 0$ or x^2 is in P.*

Proof. Let x be an element of our ordered field. By the trichotomy axiom (OF1), either $x = 0$, x is in P, or $-x$ is in P. Thus we consider three cases.

Suppose first that $x = 0$. Precisely the same argument used to prove Proposition 1.1 shows that $x^2 = 0 \cdot 0 = 0$.

Next, suppose x is in P. Then by the closure of the set of positives under multiplication (OF3), $x^2 = x \cdot x$ is in P.

Finally, suppose that $-x$ is in P. Thus by the closure of the set of positives under multiplication, $(-x)^2 = (-x) \cdot (-x)$ is in P. By the same argument that proves part (d) of Exercise 1.1, $(-x)^2 = x^2$. Thus x^2 is in P. Thus in all cases, either $x^2 = 0$ or x^2 is positive, as claimed. □

Exercise 1.6. Let F be an ordered field.

(a) Prove that 1, the multiplicative identity, must be in P and that -1 can not be in P.

(b) Prove that, in an ordered field, there is no element x for which $x^2 = -1$. (For those of you who know something about the complex numbers $\mathbb{C}$, this exercise shows that there is no order relation on $\mathbb{C}$ making it an *ordered* field.)

Again, we don't wish to devote excessive time to proving that all the usual manipulations we do with inequalities are valid. At this point we simply want you to be aware of the defining properties of an order relation and the fact that properties of an ordered field justify the familiar algebraic manipulations we do with inequalities. We do, however, want to present several very important inequalities that we will use extensively throughout the book, especially in the second half. These inequalities involve the absolute value of a real number. Recall,

$$|x| = \begin{cases} x & x \geq 0 \\ -x & x < 0 \end{cases}.$$

Proposition 1.4. *For all x, y in $\mathbb{R}$,*

(a) $|xy| = |x||y|$,

(b) $x^2 = |x|^2$,

(c) $x \leq |x|$.

Exercise 1.7. Prove Proposition 1.4.

Proposition 1.5. *For all x, y in $\mathbb{R}$,*

(a) (Triangle Inequality) $|x + y| \leq |x| + |y|$,

(b) (Reverse Triangle Inequality) $|x - y| \geq ||x| - |y||$.

Proof. We first prove part (a). Observe that

$$\begin{aligned} |x+y|^2 &= (x+y)^2 && \text{(by part (b) of Proposition 1.4)} \\ &= x^2 + 2xy + y^2 \\ &= |x|^2 + 2xy + |y|^2 && \text{(by part (b) of Proposition 1.4)} \\ &\leq |x|^2 + 2|xy| + y^2 && \text{(by part (c) of Proposition 1.4)} \\ &= |x|^2 + 2|x||y| + |y|^2 && \text{(by part (a) of Proposition 1.4)} \\ &= (|x| + |y|)^2. \end{aligned}$$

Because $|x + y|$ and $|x| + |y|$ are both non-negative, we may conclude that $|x + y| \leq |x| + |y|$, as claimed.

Next we prove part (b). We write $x = (x - y) + y$ and apply the triangle inequality to obtain

$$|x| = |(x - y) + y| \leq |x - y| + |y|.$$

Rearranging gives $|x - y| \geq |x| - |y|$. We could have started with $y = (y - x) + x$ instead and obtained

$$|x - y| \geq |y| - |x| = -(|x| - |y|).$$

Because $|x - y| \geq |x| - |y|$ and $|x - y| \geq -(|x| - |y|)$ both hold, we conclude that $|x - y| \geq ||x| - |y||$, as claimed. □

We end this section with a fact about the ordered field of real numbers that will be used surprisingly often. The result will strike you as obvious; the proof requires one to understand the rigorous definition of the real number system. Because we are postponing the discussion of the construction of the real number system to the second half of the book, at this point we treat the result as a basic fact.

Proposition 1.6 (Archimedean Property). *Let $x \in \mathbb{R}$ with $x > 0$. Then there exists $N \in \mathbb{N}$ such that $x > \frac{1}{N}$.*

The rational number system also satisfies the Archimedean property, though this is not a particularly deep result; if x is a positive rational number, we can write $x = \frac{m}{n}$ for natural numbers m and n. If $m > 1$, $\frac{m}{n} > \frac{1}{n}$. If $m = 1$,

$$\frac{m}{n} = \frac{1}{n} = \frac{2}{2n} > \frac{1}{2n}.$$

In either case, we have obtained a natural number with the desired property.

1.3 A First Look at Sets and Functions

1.3.1 Sets, elements, and subsets

In mathematics, in order to avoid circular definitions, we must agree to leave a small number of terms undefined. The notion of *set* is one such undefined term. Let S be a set. We use the notation $x \in S$ to mean that *x is an element of S* and $x \notin S$ to mean that x is not an element of S. If S has no elements, we say S is *empty* and write $S = \emptyset$. We say that *T is a subset of S* and write $T \subseteq S$, if every element of T is also an element of S. If T is a subset of S and T is not equal to S, we call T a *proper* subset of S and write $T \subset S$.

We claim that, for any set S, $\emptyset \subseteq S$. Such a claim may initially strike you as strange, but it follows from the definition; it is indeed true that every element of the empty set (for there are none) is in S. Such a statement is often said to be *vacuously true*. When my daughter was a small child, we used to have great fun with vacuously true statements like "every dragon living under my bed likes cheesy potato chips."

Example 1.2. The elements of a set may be numbers, but they could be anything really, such as functions, matrices, other sets, etc. For example, consider $S = \{0, 1\}$. We can form a new set $\mathcal{P}(S)$, called the *power set* of S, consisting of all subsets of S. Thus

$$\mathcal{P}(S) = \{\emptyset, \{0\}, \{1\}, \{0, 1\}\}.$$

Exercise 1.8. Carefully explain the difference between the two sets $\emptyset$ and $\{\emptyset\}$.

Exercise 1.9. If $T = \{a, b, c\}$, find $\mathcal{P}(T)$, the set of all subsets of T.

Rather than listing the elements of a set, we can define a set by specifying the identifying property the elements satisfy. For example,

$$S = \{n \in \mathbb{Z} : 0 \leq n \leq 3\}$$

is an alternate way to describe the set $\{0, 1, 2, 3\}$.

1.3.2 Operations with sets

Given two sets S and T, there are several ways we can combine them to obtain a new set. We define

$$\begin{aligned} S \cup T &= \{x : x \in S \text{ or } x \in T\} \\ S \cap T &= \{x : x \in S \text{ and } x \in T\} \\ T^c &= \{x : x \notin T\} \\ S \setminus T &= \{x : x \in S \text{ and } x \in T^c\}. \end{aligned}$$

Note that the *or* we use in math is the *inclusive or*; thus we interpret the definition of $S \cup T$ above to mean that elements in both S and T are included in $S \cup T$. $S \cup T$ is called the *union* of S and T and $S \cap T$ is called the *intersection* of S and T. T^c is called the *complement* of T.

1.3.3 Special subsets of $\mathbb{R}$: intervals

For certain kinds of sets that we encounter frequently, it is convenient to have simpler notation. Thus we define *intervals* in $\mathbb{R}$.

Definition 1.5. Let $a, b \in \mathbb{R}$ with $a < b$. Then we define the **open interval** (a, b) to be

$$(a, b) = \{x \in \mathbb{R} : a < x < b\}$$

and we define the **closed interval** $[a, b]$ to be

$$[a, b] = \{x \in \mathbb{R} : a \leq x \leq b\}.$$

We also define half-open intervals:

$$[a, b) = \{x \in \mathbb{R} : a \leq x < b\}$$

and

$$(a, b] = \{x \in \mathbb{R} : a < x \leq b\}.$$

We also allow a or b to be infinite, writing (a, ∞) for $\{x \in \mathbb{R} : x > a\}$, $(-\infty, b]$ for $\{x \in \mathbb{R} : x \leq b\}$, etc.

Example 1.3. Let $A = (0, 2)$ and $B = [1, 3]$. We want to describe $A \cup B$ and $A \cap B$.

Let x denote a real number. $x \in A$ if

$$0 < x < 2,$$

and $x \in B$ if

$$1 \leq x \leq 3.$$

The set $A \cup B$ is the set of x satisfying either inequality. Such x must satisfy

$$0 < x \leq 3.$$

Thus $A \cup B$ is the half-open interval $(0, 3]$.

The set $A \cap B$ is the set of all x satisfying both inequalities above. Such x satisfy

$$1 \leq x < 2.$$

Thus $A \cap B = [1, 2)$.

Exercise 1.10. Let $c, d \in \mathbb{R}$.

(a) If $c < d$, find $(-\infty, c) \cap (d, \infty)$.

(b) If $c > d$, find $(-\infty, c) \cap (d, \infty)$ and $(-\infty, c) \cup (d, \infty)$.

(c) If $c = d$, find $\mathbb{R} \setminus ((-\infty, c) \cup (d, \infty))$.

1.3.4 Functions

We end this chapter by reviewing the basics concerning functions.

Definition 1.6. Let A and B be non-empty sets. A **function** $f\colon A \to B$ is a rule that assigns to each element a in A a unique element $f(a)$ in B. We call A the **domain**, B the **range**, and $f(a)$ the **image of** a **under** f.

We note that although the definition requires *every* element of A to have an image $f(a)$ in B, it *does not require* every element of B to be the image of an element of A. We denote by $f(A)$ that subset of B consisting of images of elements of A.

Example 1.4. Consider $f\colon \mathbb{R} \setminus \{0\} \to \mathbb{R}$ defined by the equation $f(x) = \frac{1}{x} + 1$. Here, the domain is $\mathbb{R} \setminus \{0\}$; we do not specify a value for the function at 0. In this case, we have said that the range of f is $\mathbb{R}$, and indeed it is true that for every $x \neq 0$, $\frac{1}{x} + 1$ is an element of $\mathbb{R}$. We claim that $f(\mathbb{R} \setminus \{0\}) = \mathbb{R} \setminus \{1\}$.

Indeed, we note that 1 is not in the image of f, for if it were, there would be an $x \in \mathbb{R}$ satisfying

$$\frac{1}{x} + 1 = 1,$$

or, equivalently,

$$\frac{1}{x} = 0.$$

No such x exists.

We claim, however, that if $b \neq 1$, there exists $x \in \mathbb{R}$ such that $f(x) = b$. Such an x would satisfy

$$\frac{1}{x} + 1 = b.$$

Solving for x (recalling that $b \neq 1$) gives

$$x = \frac{1}{b-1}.$$

Thus we have shown that every b in $\mathbb{R} \setminus \{1\}$ is in the image of f.

Exercise 1.11. Explain carefully why there is no real, non-zero x satisfying $\frac{1}{x} = 0$.

1.4 Problems

1. Let $3\mathbb{Z}$ denote the set of all integers divisible by 3.

   (a) Show that $3\mathbb{Z}$ is closed under the usual operations of addition and multiplication.

   (b) Consider the five properties satisfied by $\mathbb{Z}$ with its standard operations of addition and multiplication. Determine whether $3\mathbb{Z}$ with its standard operations of addition and multiplication satisfies these properties.

2. Determine whether each of the following is true or false for $x, y, z \in \mathbb{R}$. Prove each true statement and provide a counterexample for each false statement.

   (a) $x \leq x^2$ for all $x \in \mathbb{R}$.

   (b) If $x < y$, then $xz < yz$.

   (c) $x^3 \leq x^3 + y^3$ for all $x, y \in \mathbb{R}$.

3. Study the proof of the triangle inequality. Under what conditions (if any) on x and y is it true that $|x + y| = |x| + |y|$?

4. Consider $f\colon \mathbb{R} \to \mathbb{R}$ given by $f(x) = \frac{1}{1+x^2}$. Find $f(\mathbb{R})$, the image of $\mathbb{R}$ under f. As in Example 1.4, you should demonstrate that $f(\mathbb{R})$ is as you claim by showing that any point outside your set is not in the image of f and any point in your set is in the image of f.

Chapter 2

Mathematical Induction

2.1 First Examples

Suppose we want to find a simple formula for the sum of the first n odd numbers:

$$1 + 3 + 5 + \ldots + (2n - 1) = \sum_{k=1}^{n} (2k - 1).$$

How might we proceed? The most natural approach is to do the calculation for several small values of n. Doing so, we find

$$\begin{aligned}
n = 1: &\qquad \sum_{k=1}^{1} (2k - 1) = 1 \\
n = 2: &\qquad \sum_{k=1}^{2} (2k - 1) = 1 + 3 = 4 \\
n = 3: &\qquad \sum_{k=1}^{3} (2k - 1) = 1 + 3 + 5 = 9 \\
n = 4: &\qquad \sum_{k=1}^{4} (2k - 1) = 1 + 3 + 5 + 7 = 16.
\end{aligned}$$

These examples lead us to the conjecture

$$\sum_{k=1}^{n} (2k - 1) = n^2. \tag{2.1}$$

Is our conjecture correct? It certainly agrees with the results obtained for $n = 1, 2, 3, 4$ above, but how do we know whether it is true *for all* n? No amount of computing will answer this question; what we need is a general argument. That is, we need a proof.

Here is one elementary argument: Let $S[n] = \sum_{k=1}^{n}(2k-1)$. Then the following sum is equal to $2S[n]$:

$$\begin{array}{ccccccccc} 1 & + & 3 & + & \dots & + & (2n-3) & + & (2n-1) \\ (2n-1) & + & (2n-3) & + & \dots & + & 3 & + & 1 \end{array}.$$

We have n columns with two numbers each, and each column sums to $2n$. Thus

$$2S[n] = n \cdot 2n = 2n^2,$$

and hence $S[n] = n^2$.

This example illustrates a common situation in which we want to establish an *infinite* number of statements, one for each natural number n. In the above example, a bit of clever grouping established the general result. Not surprisingly, the number of situations in which such an argument is possible is small. The principle of mathematical induction will give us a general approach to obtaining proofs of statements like the above. We will discuss this general principle later in the section, once we have established some of the concepts necessary to formulate and explore a wider range of questions.

2.1.1 Defining sequences through a formula for the n-th term

Definition 2.1. A **sequence** a of real numbers is a function from the set $\mathbb{N}_0$ of **non-negative integers** $0, 1, 2, 3, \dots$ to the set $\mathbb{R}$ of real numbers. We write either $a[n]$ or a_n for the value of the function at n and call this value the n-th term of the sequence. We denote the entire sequence by a or $\{a_n\}$.

Note that $a[n]$ and a_n are two different notations for the n-th term of the sequence a. We will use them interchangeably (though we will be consistent within any example or proposition). The former notation may help you remember that a is a function, and that each term is really the value of this function at one of the elements of the domain. It is also the notation we'll use when we write Python programs. The latter notation is the traditional notation for terms in a sequence and will be used when we are not developing a program.

Remark 1 (zero indexing). Some books define a sequence a to be a function whose domain is the set $\mathbb{N}$ of natural numbers $1, 2, 3, \dots$. Perhaps this definition seems more natural, for then the first five terms are a_1, a_2, a_3, a_4, and a_5 rather than a_0, a_1, a_2, a_3, and a_4. Our reason for starting at zero is that we will be programming in Python, and Python (and many other programming languages) use zero-indexed sequences. We want to make the notation in our examples compatible with the syntax we will use in our programs.

Were we British, zero indexing might seem more natural; in a multi-story building, the ground floor is the zeroth floor and the first floor is the next one up.

Example 2.1. In the previous section, we considered the sequence S, where, for $n \geq 1$, $S[n]$ is the sum of the n odd numbers $1, 3, \ldots, 2n-1$. It is reasonable to define $S[0]$ to be the empty sum, so that $S[0] = 0$. We showed above that $S[n] = n^2$ for $n \geq 1$. This formula is also valid for $n = 0$, and thus we have a simple formula for the n-th term of the sequence S.

Sometimes, as above, it is rather easy to guess a formula for the n-th term of a sequence by examining the first few terms. What can we try when it is not so obvious? It is sometimes helpful to look at the sequence of *ratios* or *differences* of consecutive terms, or the sequence of differences of consecutive terms in the sequences of differences, etc. We illustrate with some examples.

Constant differences: arithmetic sequences. Consider the sequence $\{a_n\}$ of all natural numbers for which the remainder when divided by 10 is 3. The first few terms of the sequence are

$$a_0 = 3, \quad a_1 = 13, \quad a_2 = 23, \quad a_3 = 33.$$

Consider the sequence $\{d_n\}$ of differences of consecutive terms. Thus $d_0 = a_1 - a_0 = 13 - 3 = 10$, $d_1 = a_2 - a_1 = 23 - 13 = 10$, $d_3 = a_4 - a_3 = 33 - 23 = 10$, and in general $d_n = a_{n+1} - a_n = 10$. The sequence $\{a_n\}$ is an example of an *arithmetic sequence*, that is, a sequence for which the sequence $\{d_n\}$ of differences is constant. One can show (see the next exercise) that the n-th term formula for a sequence $\{a_n\}$ is linear in n if and only if the sequence $\{d_n\}$ of differences is constant. In this example,

$$\begin{aligned}
a_0 &= 3 \\
a_1 &= 13 = 3 + 10 \\
a_2 &= 23 = 13 + 10 = 3 + 10 + 10 = 3 + 10 \cdot 2 \\
a_3 &= 33 = 23 + 10 = 3 + 2 \cdot 10 + 10 = 3 + 10 \cdot 3 \\
&\vdots \\
a_n &= 3 + 10n.
\end{aligned}$$

Exercise 2.1. Consider the statement: The n-th term formula for a sequence $\{a_n\}$ is linear in n if and only if the sequence $\{d_n\}$ of differences is constant. This fact should be familiar to you even if the phrasing in terms of sequences is not. Reconcile the above statement with what you know about linear functions from algebra.

Constant ratios: geometric sequences. Consider the sequence whose terms are the successive powers of 2, beginning with $2^0 = 1$. The first few terms of the sequence are

$$\begin{aligned}a_0 &= 2^0 = 1\\ a_1 &= 2^1 = 2\\ a_2 &= 2^2 = 4\\ a_3 &= 2^3 = 8,\end{aligned}$$

and, in general, $a_n = 2^n$. This sequence is not arithmetic; the differences of consecutive terms are

$$\begin{aligned}d_0 &= a_1 - a_0 = 2 - 1 = 1\\ d_1 &= a_2 - a_1 = 4 - 2 = 2.\end{aligned}$$

Let's look at *ratios* of consecutive terms:

$$\begin{aligned}r_0 &= \frac{a_1}{a_0} = \frac{2}{1} = 2\\ r_1 &= \frac{a_2}{a_1} = \frac{4}{2} = 2\\ r_2 &= \frac{a_3}{a_2} = \frac{8}{4} = 2\\ r_n &= \frac{a_{n+1}}{a_n} = \frac{2^{n+1}}{2^n} = 2.\end{aligned}$$

Such a sequence, in which the sequence $\{r_n\}$ of ratios of consecutive terms is constant, is called a *geometric sequence.* When r_n is equal to some constant r for all n, we call r the *common ratio.*

Exercise 2.2. Determine whether each sequence is arithmetic, geometric, or neither and find a formula for the n-th term.

(a) A ball is dropped from a height of 2 meters. Each time it bounces, it achieves a height that is $\frac{1}{3}$ of its height on the preceding bounce. Let $h_0 = 2$ and let h_n be the height achieved after the n-th bounce.

(b) Let S_n be the sum of the first n natural numbers that are multiples of 3. Interpret S_0 as the empty sum, with value 0.

(c) A car rental company charges a flat rate of \$50 plus \$0.60 for each mile traveled. Let C_n be the total car rental bill if the car is driven n miles.

2.1.2 Defining sequences recursively

We begin with an example. For $n \in \mathbb{N}_0$, the *n-th triangular number* T_n is the number of dots in a triangular array of dots in which there are n rows and the

j-th row has j dots. Thus $T_0 = 0$ (for we have no rows of dots) and T_1, T_2, and T_3 are shown below:

Thus the first few triangular numbers are

$$T_0 = 0, \quad T_1 = 1, \quad T_2 = 3, \quad T_3 = 6, \quad T_4 = 10.$$

We can define the sequence for arbitrary n as follows: $T_0 = 0$, and for $n \geq 1$, $T_n = T_{n-1} + n$. In such a *recursive definition* for a sequence, the n-th term is defined to be a function of one or more of the terms that precede it. With a recursive definition, one must also separately specify the first few terms; if the n-th term is a function of the preceding k terms, one must specify the first k terms of the sequences.

Exercise 2.3. Write a recursive definition for the arithmetic sequence

$$\{3, 13, 23, 33, \ldots\}$$

considered above.

Exercise 2.4. Write a recursive definition for the geometric sequence

$$\{1, 2, 4, 8, \ldots\}$$

considered above.

Let's return to our exploration of the triangular numbers. Although we have a recursive definition, we would still like an explicit formula for the n-th term; indeed, with a recursive definition, if we want to know, say, T_{1000}, we must first generate $T_1, T_2, \ldots, T_{999}$. One can obtain an n-th term formula in a number of ways. One straight-forward way involves *iterating* the recursive formula for T_n:

$$\begin{aligned} T_n &= T_{n-1} + n \\ &= T_{n-2} + (n-1) + n \\ &= T_{n-3} + (n-2) + (n-1) + n \\ &= 0 + 1 + 2 + \ldots + (n-2) + (n-1) + n \\ &= \sum_{j=0}^{n} j. \end{aligned}$$

We can now obtain a simple formula for T_n in the same way we obtained a formula for the sum of the first n odd numbers. We find $T_n = \frac{n(n+1)}{2}$.

Exercise 2.5. Verify the last assertion.

Exercise 2.6. Give a "picture proof" of the formula $T_n = \frac{n(n+1)}{2}$ by considering an n by $n+1$ rectangular array of dots.

Exercise 2.7. Suppose a sequence is defined recursively by $a_0 = 4$ and $a_n = na_{n-1}$ for $n \geq 1$. Use iteration to obtain an n-th term formula for a_n.

We consider another approach to obtaining an n-th term formula for T_n; as we did with the arithmetic sequence in the last subsection, let us look at the sequence $\{d_n\}$ of differences of consecutive terms.

$$d_n = T_{n+1} - T_n = (T_n + n + 1) - T_n = n + 1.$$

This sequence of differences is *not* constant. However, we may consider the sequence $\{d_n^{(2)}\}$ of differences for the sequence $\{d_n\}$. We find

$$d_n^{(2)} = d_{n+1} - d_n = (n+2) - (n+1) = 1.$$

In other words, the sequence of *second differences* is now constant. We saw above that when the sequence of *first* differences of a is constant, then the n-th term of a is a linear function of n. Suppose we knew that whenever the sequence of *second* differences is constant, the n-th term of a is a quadratic function of n. We would then have

$$T_n = An^2 + Bn + C$$

for constants A, B, and C to be determined. We can solve for these constants using the first three triangular numbers. We obtain

$$\begin{aligned} 0 &= T_0 = C \\ 1 &= T_1 = A + B + C \\ 3 &= T_2 = 4A + 2B + C. \end{aligned}$$

Solving this system gives $A = B = \frac{1}{2}$ and $C = 0$, i.e.,

$$T_n = \frac{1}{2}n^2 + \frac{1}{2}n.$$

We obtained the same formula above. *It remains to prove that, when the sequence of second differences is constant, the n-th term of the original sequence is a quadratic function of n.* We leave the proof of this claim to the reader.

Exercise 2.8. (Challenging) Show that the sequence $\{d_n^{(2)}\}$ of second differences of $\{a_n\}$ is a non-zero constant sequence if and only if $a_n = q(n)$ for some quadratic function q. In other words, you must prove two things:

(i) If there are constants A, B, and C with A non-zero such that

$$a_n = An^2 + Bn + C \quad \text{for all } n \geq 0,$$

then the sequence of second differences is a non-zero constant sequence.

(ii) If the sequence of second differences is a non-zero constant sequence, then there exists a quadratic function q such that $a_n = q(n)$ for all n.

2.2 First Programs

So far, our sequences have been simple enough that we could do all our calculations by hand. As our examples become more complicated, calculating by hand may become tedious or impossible. We therefore put our technology to use.

Let's look for a simple formula for

$$S[n] = \sum_{j=1}^{n} j^2 = 1^2 + 2^2 + 3^2 + \ldots + n^2.$$

We would like our sequence to be zero-indexed (see Remark 1 above), and so we let $S[0] = 0$. This definition is consistent with our definition of $S[n]$ for $n > 0$ because, for $n = 0$, the sum is empty. Note that our sequence has the simple recursive definition

$$S[n] = S[n-1] + n^2, \quad n \geq 1.$$

Inspired by the example of the triangular numbers, we will generate the first terms of the sequence S and then look at the sequence $d^{(1)}$ of first differences, the sequence $d^{(2)}$ of second differences, and so on. Our goal is to figure out whether there is an m so that the sequence $d^{(m)}$ of m-th differences is constant. Our experience suggests that if such an m exists, then the n-th term formula for S will be $S[n] = p(n)$ where p is a polynomial of degree m. We will still have work to do, for we have only *proved* this statement for $m = 1$ and $m = 2$. But in the next section we will see a powerful technique that will allow us to prove a conjectured formula for $S[n]$ no matter how it was obtained.

We use this example to develop our first Python program. If you are new to Python, you should first read the appendix.

The first thing the program should do is produce a number of terms in the sequence S. Let's generate the first 15 terms, $S[0], S[1], \ldots, S[14]$. We can use a simple loop. As above, we will name our list S. Because this sequence is defined recursively, we seed the list with the first term, 0. We now go through a loop 14 times (for $1 \leq n < 15$), each time replacing our existing list with the list obtained by appending the next term. Here's the code:

```
N = 15 #number of terms we'll generate
S = [0]#The first term in the sequence is 0
for n in range(1,N):
    S.append(S[n-1]+n**2)
```

Next, we must generate terms in the sequence $d^{(1)}$ of first differences. Recall that $d^{(1)}[0] = S[1] - S[0], \ldots, d^{(1)}[13] = S[14] - S[13]$. Thus because we only generated 15 terms of S, we may only generate the first 14 terms of $d^{(1)}$. We do not need to seed the list $d^{(1)}$, so we may initially define it to be an empty list.

```
d1 = []
for n in range(1,len(S)):
    d1.append(S[n]-S[n-1])
```

Similar code generates the first 13 second differences and the first 12 third differences. The full program appears below, with the output included in a comment.

```
N =15 #number of terms we'll generate

S = [0]
for n in range(1, N):
    S.append( S[n-1] + n**2 )

d1 = []
for n in range(1, len(S)):
    d1.append( S[n] - S[n-1] )

d2 = []
for n in range(1, len(d1)):
    d2.append( d1[n] - d1[n-1] )

d3 = []
for n in range(1, len(d2)):
    d3.append( d2[n] - d2[n-1] )

print("Sequence: " + str(S))
print("First Differences: " + str(d1))
print("Second Differences: " + str(d2))
print("Third Differences: " + str(d3))

#output
#Sequence: [0, 1, 5, 14, 30, 55, 91, 140, 204, 285, 385, 506,
    650, 819, 1015]
#First Differences: [1, 4, 9, 16, 25, 36, 49, 64, 81, 100, 121,
    144, 169, 196]
#Second Differences: [3, 5, 7, 9, 11, 13, 15, 17, 19, 21, 23,
    25, 27]
#Third Differences: [2, 2, 2, 2, 2, 2, 2, 2, 2, 2, 2, 2]
```

The output of the program suggests (but does not prove) that the sequence $d^{(3)}$ is constant. Our experience thus far leads us to conjecture that the n-th term of the sequence S is given by $S[n] = An^3 + Bn^2 + Cn + D$ for appropriate constants A, B, C, and D.

Exercise 2.9. As above, obtain a linear system for A, B, C, and D and solve it to obtain

$$S[n] = \frac{1}{3}n^3 + \frac{1}{2}n^2 + \frac{1}{6}n = \frac{n(2n^2 + 3n + 1)}{6} = \frac{n(n+1)(2n+1)}{6}. \tag{2.2}$$

Exercise 2.10. Modify the code to generate the sequence T of triangular numbers from the previous subsection, together with its sequences of first, second, and third differences.

2.3 First Proofs: The Principle of Mathematical Induction

Our work in the previous section leads to the conjecture:

$$S_n = \sum_{j=1}^{n} j^2 = \frac{n(n+1)(2n+1)}{6} \quad \text{for all } n \in \mathbb{N}.$$

This formula agrees with the output of our program. How can we know that the formula is valid *for all* n? No amount of computing can settle this question; no matter how long we run our program, we will have only checked our result for finitely many n. What we need is a proof. Although many arguments are possible, we give a proof using the principle of mathematical induction.

Principle of Mathematical Induction. Suppose we have an infinite collection of statements $P(1), P(2), \ldots, P(n), \ldots$, one for each natural number n. Suppose we know

(i) $P(1)$ is true, and

(ii) whenever $P(k)$ is true, then $P(k+1)$ is true.

Then $P(n)$ is true for all n.

In our example, the statement $P(n)$ is "S_n, the sum of the first n perfect squares, equals $\frac{n(n+1)(2n+1)}{6}$." Consider the *base case*, i.e., the statement when $n = 1$. Clearly $S_1 = 1$. Because

$$\frac{1(1+1)(2(1)+1)}{6} = \frac{1 \cdot 2 \cdot 3}{6} = 1,$$

it is indeed the case that $S_1 = \frac{1(1+1)(2(1)+1)}{6}$. Thus $P(1)$ is true.

For the *induction step*, we suppose that $P(k)$ is true for some $k \geq 1$ and we consider $P(k+1)$. In other words, we consider

$$S_{k+1} = \sum_{j=1}^{k+1} j^2 = \left(\sum_{j=1}^{k} j^2 \right) + (k+1)^2.$$

By the induction hypothesis, $P(k)$ is true, and so the sum of the first k squares can be replaced by $\frac{k(k+1)(2k+1)}{6}$. We find

$$\begin{aligned} S_{k+1} &= \frac{k(k+1)(2k+1)}{6} + (k+1)^2 \\ &= \frac{(k+1)}{6}\left[k(2k+1) + 6(k+1)\right] \\ &= \frac{(k+1)}{6}\left[2k^2 + 7k + 6\right] \\ &= \frac{(k+1)}{6}\left[(k+2)(2k+3)\right] \\ &= \frac{(k+1)[(k+1)+1][2(k+1)+1]}{6}. \end{aligned}$$

Thus $P(k+1)$ is true if $P(k)$ is true. By the principle of mathematical induction, $P(n)$ is true for all n. □

You may have noticed that we did not *prove* that the principle of mathematical induction is indeed a valid proof technique. The full justification requires us to give a more rigorous treatment of the natural numbers than we did in the first chapter. Such a treatment is beyond the scope of this course. We note, however, that if we accept the principle of mathematical induction, we can use it to prove a useful generalization:

Proposition 2.1. *Consider a collection of statements $P(a), P(a+1), P(a+2), \ldots$ for some integer a. Suppose (i) $P(a)$ is true and (ii) whenever $P(k)$ is true for some integer k, $P(k+1)$ is true. Then $P(n)$ is true for all $n \geq a$.*

We leave the proof to the reader.

Exercise 2.11. Prove Proposition 2.1.

The point of the proposition is that we may use the technique of induction even when the base case is some integer other than 1.

In the above example of summing the squares, we were a bit more verbose than we will tend to be in the future. For example, we will seldom say explicitly what the statement $P(n)$ is because it is usually obvious. We illustrate by proving another proposition by induction. Note also that our base case will be $n = 0$ and not $n = 1$.

Proposition 2.2. *(Bernoulli's inequality) Let $x \geq -1$. Then for every integer $n \geq 0$,*

$$(1+x)^n \geq 1 + nx. \tag{2.3}$$

Proof. The proof is by induction on n with base case $n = 0$. When $n = 0$,

$$(1+x)^0 = 1 = 1 + 0x$$

and the result holds.

Now suppose the result holds for some $k \geq 0$ and consider $k+1$. Then

$$(1+x)^{k+1} = (1+x)(1+x)^k.$$

By the induction hypothesis, $(1+x)^k \geq 1+kx$. Also, because $x \geq -1$, $1+x \geq 0$. Thus

$$\begin{aligned}(1+x)^{k+1} &= (1+x)(1+x)^k \\ &\geq (1+x)(1+kx) \\ &= 1+kx+x+kx^2 \\ &\geq 1+(k+1)x+0,\end{aligned}$$

and the result indeed holds for $k+1$. By the principle of mathematical induction, it holds for all integers $n \geq 0$. □

Exercise 2.12. Prove by induction that $4^n - 1$ is a multiple of 3 for all $n \geq 0$.

2.4 Strong Induction

Some proofs require the *strong principle of mathematical induction*:

Strong Principle of Mathematical Induction. Suppose we have an infinite collection of statements $P(n)$, one for each $n \in \mathbb{N}$. Suppose

(i) $P(1)$ is true, and

(ii) for $k \geq 1$, if $P(i)$ is true for all $1 \leq i \leq k$, then $P(k+1)$ is true.

Then $P(n)$ is true for all n.

Although it seems as if we are assuming more at the induction step, the strong principle of mathematical induction and the usual principle of mathematical induction are, in fact, equivalent. We give an example of a situation in which we need this stronger induction principle

Example 2.2. Suppose we have 3-cent and 5-cent stamps. What postage can be made?

Clearly we can not make postage of 1, 2, 4, or 7 cents. It appears that we can make all the others. Indeed, postages of 3 and 5 cents are trivial, and

$$\begin{aligned}6 &= 3+3 \\ 8 &= 5+3 \\ 9 &= 3+3+3 \\ 10 &= 5+5 \\ 11 &= 5+3+3,\end{aligned}$$

and so on. It thus appears that any postage of 8 cents or more can be made. We prove this conjecture using the strong induction principle.

We have just shown explicitly how to make postage of 8, 9, 10, or 11 cents. Suppose now that $k \geq 11$ and that, for all $8 \leq i \leq k$, we can make postage of

i cents using 3- and 5-cent stamps. Consider postage of $k+1$ cents. Because $k \geq 11$, $(k+1)-3 = k-2 \geq 9$. By the strong induction hypothesis, we can make postage of $k-2$ cents, and by adding an additional 3-cent stamp, we can make postage of $k+1$ cents. Thus by the strong principle of mathematical induction, we can make any postage of 8 cents or more with 3- and 5-cent stamps.

Exercise 2.13. Determine all postage that can be made using 4- and 7-cent stamps.

2.5 The Well-Ordering Principle and Induction

The principle of mathematical induction can be thought of as a statement about the natural numbers. It essentially says that the natural numbers are the smallest set with the property that 1 is in the set, and whenever k is in the set, $k+1$ is in the set. The principle of mathematical induction turns out to be equivalent to another statement about the natural numbers called the well-ordering principle.

Well-Ordering Principle. Every non-empty subset S of $\mathbb{N}$ has a smallest element.

Proposition 2.3. *The well-ordering principle and the strong principle of mathematical induction are equivalent.*

Proof. First we assume that the strong principle of mathematical induction holds and we prove the well-ordering principle. Thus suppose S is a non-empty subset of $\mathbb{N}$. Let $T = \mathbb{N} \setminus S$. Suppose, for a contradiction, that S does not have a smallest element. Then $1 \notin S$, for if it were, it would be the smallest element of S because it is the smallest element of $\mathbb{N}$. Thus $1 \in T$.

Now suppose for some k, for all $1 \leq i \leq k$, $i \in T$. Consider $k+1$. If $k+1$ were in S, it would be the smallest element of S because no i between 1 and k is in S. Because we are assuming S does not have a smallest element, we must have $k+1 \in T$. Then, by the strong principle of mathematical induction, T contains all natural numbers and is hence equal to $\mathbb{N}$. It follows that $S = \emptyset$. We have reached a contradiction and so we conclude that S has a smallest element.

Next we suppose that the well-ordering principle holds and we use it to prove the strong principle of mathematical induction. Thus we suppose we have a collection of statements $P(n)$, one for each natural number n. We suppose $P(1)$ is true and that whenever $P(i)$ is true for all $1 \leq i \leq k$, $P(k+1)$ is true. Let T denote the set of all natural numbers n for which $P(n)$ is *not* true. Our goal is to show that T is empty, for then we can conclude that $P(n)$ is true for all n.

Suppose, for a contradiction, that T is not empty. Then by the well-ordering principle, T has a smallest element m. m can not equal 1 because $P(1)$ is true. Thus $m-1$ is still an element of $\mathbb{N}$. Now, because m is the smallest natural number for which $P(m)$ is not true, for all $1 \leq i \leq m-1$, $P(i)$ is true. By hypothesis, $P((m-1)+1) = P(m)$ is true and $m \notin T$. This contradiction proves that T is empty, as desired. □

Exercise 2.14. The purpose of this exercise is to exhibit an example of a set of real numbers for which the well-ordering principle does not hold.

(a) Give an example of a subset of $\mathbb{Q}$ that has a smallest element.

(b) Give an example of a subset of $\mathbb{Q}$ that does *not* have a smallest element.

2.6 Problems

Whenever possible, we pose problems in which you need to explore, conjecture, and prove. You may, of course, explore however you like, using trial and error, doing concrete examples via paper-and-pencil calculations, or using your developing programming skills to write a short program to do many examples quickly for you. There is no "right" way to explore. Needless to say, most of the problems posed ask you to discover a known result. Thus were you to open other books or go online, you could find formulas and proofs. The point, of course, is to learn how to discover results for yourself.

A final note: Just because the title of the chapter is *Mathematical Induction* does not mean you must or should use induction to do all of these problems. For many of them, induction will work. But for many, other proofs exist as well, and you should look for them. Each new proof gives new insight.

1. Find and prove a formula for the sum of the first n even natural numbers.

2. Find and prove a formula for the sum of the first n perfect cubes, i.e., for $\sum_{j=1}^{n} j^3$.

3. Let $a_n = 11^n - 6$ for $n \geq 1$. By writing down the prime factorizations of the first few terms, make a conjecture about a factor they all have in common. Prove your claim.

4. Find and prove a formula for the sum of the first n powers of 2, beginning with 2^0. That is, find and prove a formula for $\sum_{j=0}^{n-1} 2^j = 2^0 + 2^1 + 2^2 + \ldots + 2^{n-1}$. See if you can give more than one proof.

5. Find and prove a formula for $\sum_{j=0}^{n-1} 3^j$, the sum of the first n powers of 3 beginning with 3^0.

6. A *finite geometric series* is a series of the form $\sum_{j=0}^{n-1} ar^j$ for non-zero a and r. Show that, for $r \neq 1$,

$$\sum_{j=0}^{n-1} ar^j = \frac{a(1-r^n)}{1-r}.$$

7. Consider the sequence $\{a_n\}$ defined recursively by $a_0 = 1$ and $a_n = a_{n-1} + 4$ for $n \geq 1$. Find an explicit formula for a_n and an explicit formula for $\sum_{j=0}^{n} a_j$.

8. Recall that a sequence $\{c_n\}$ is called *arithmetic* if it is given recursively by $c_0 = c$ and $c_n = c_{n-1} + d$ for $n \geq 1$ and for fixed constants c and d. Find n-th term formulas for c_n and for $S_n = \sum_{j=0}^{n} c_j$.

9. A sequence $\{a_n\}$ is *bounded above* if there exists an M such that $a_n \leq M$ for all n. Similarly, a sequence is *bounded below* if there exists m such that $a_n \geq m$ for all n. Let $a_0 = 1$ and, for $n > 0$, let $a_n = \sqrt{2a_{n-1} + 3}$.

   (a) Generate enough terms of the sequence to make a conjecture about whether or not this sequence is bounded above or below. Prove your conjecture.

   (b) A sequence $\{a_n\}$ is *non-decreasing* if, for all n, $a_n \leq a_{n+1}$. We may similarly define *non − increasing* sequences. A sequence is said to be *monotone* if it is either non-decreasing or non-increasing. Determine whether the sequence in this problem is monotone.

10. Explore the way in which the boundedness and monotonicity of the sequence in the previous problem depend on the value a_0. Give proofs of all your claims.

11. Our experience with polynomials and exponential functions suggests to us that each of the following inequalities should be true, at least for n sufficiently large. Find all n for which each inequality holds. Prove your conjecture by induction on n.

   (a) $2^n \geq 2n + 1$.

   (b) $2^n \geq n^2$.

   (c) $n! \geq 2^{n-1}$.

12. A game of Poison begins with a row of 10 pennies and two players. On each player's turn, he or she may remove one penny or two pennies. The last penny is poison, and thus the object is to force one's opponent to take the last penny. Would you prefer to be the first player or the second player? Generalize to an n-penny game, with proof, of course.

13. (Fibonacci sequence) The Fibonacci sequence $\{F_n\}$ is given recursively by $F_0 = 0$, $F_1 = 1$, and $F_n = F_{n-1} + F_{n-2}$ for $n \geq 2$.

   (a) Generate the first 20 terms of the Fibonacci sequence.

   (b) Let S_k denote the sum of the first k Fibonacci numbers with odd index. That is, $S_k = \sum_{j=1}^{k} F_{2j-1}$. Find and prove a simple formula for S_k.

(c) Let T_k denote the sum of the first k Fibonacci numbers with even index. That is, $T_k = \sum_{j=1}^{k} F_{2j-2}$. Find and prove a simple formula for T_k.

14. The purpose of this problem is to prove a number of useful inequalities.

(a) Suppose x and y are real numbers. Show that

$$2xy \leq x^2 + y^2.$$

(b) Suppose a and b are non-negative real numbers. Show that

$$\sqrt{ab} \leq \frac{a+b}{2}.$$

(This inequality is often called the *Arithmetic-Geometric Mean (AGM) inequality.*)

(c) (Challenging) Let n be a natural number, and suppose $a_k > 0$ for $1 \leq k \leq 2^n$. Prove

$$(a_1 a_2 \cdots a_{2^n})^{\frac{1}{2^n}} \leq \frac{a_1 + a_2 + \ldots + a_{2^n}}{2^n}.$$

Programming Projects.

1. In this chapter, we introduced arithmetic sequences (for which there is a common difference d between consecutive terms) and geometric sequences (for which there is a common ratio r between consecutive terms). We also saw examples of sequences for which we had to look at second and third differences in order to guess a formula for the n-th term. Write a program that takes a list consisting of the first six terms of a sequence and finds the first few terms in the sequences of first and second differences and the sequences of first and second ratios. Then use it to aid you in discovering n-th term formulas for sequences agreeing with the first six terms of each of the following:

$$a = \left\{3, \frac{3}{2}, \frac{1}{2}, \frac{1}{8}, \frac{1}{40}, \frac{1}{240}, \ldots\right\}$$
$$b = \left\{\frac{7}{3}, 3, 3, \frac{7}{3}, 1, -1, \ldots\right\}.$$

2. (Challenging) We now consider the general postage stamp problem: Suppose we have postage stamps in denominations of a cents and b cents, where $a < b$ and a and b have no common factors. What is the largest postage that can not be made? The aim of this project is to make a conjecture.

(a) Observe that postage P can be made if there exist non-negative integers n and m such that

$$P = ma + nb.$$

Write a program that takes a fixed a and b as input and generates possible values for P. Of course you can not generate all possible values of P because there are infinitely many such values. Determining sensible ranges for m and n is a hard question; we need to generate values of P beyond the largest that can not be made, but the entire purpose of this exploration is to determine the largest value of P that can not be made. The examples we have done so far have found this largest postage that can not be made to be less than ab. With this observation in mind, write your program to generate all P with $0 \leq m \leq 10b$ and $0 \leq n \leq 10a$.

(b) Now use your program to investigate how the largest P that can not be made depends on a and b. One suggestion: Begin by fixing $a = 2$ and consider $b = 3, 5, 7, 9, 11, \ldots.$ See how the largest postage that can not be made depends on b. Then start over with $a = 3$ and $b = 4, 5, 7, 8, 10, 11, \ldots.$

(c) Based on the investigations above, make a conjecture. State your conjecture in the form of a theorem to be proved. Make sure to include all hypotheses.

Chapter 3

Basic Logic and Proof Techniques

Now that we have introduced a number of mathematical objects to study and have a few proof techniques at our disposal, we pause to look a little more closely at basic logic.

3.1 Logical Statements and Truth Tables

3.1.1 Statements and their negations

Outside of mathematics, we use the word *statement* as a synonym for *declarative sentence*. Thus in ordinary English, the following are both statements.

Today is Tuesday.
Mozart is the greatest composer.

In logic and mathematics, a *statement* must have a truth value. Thus "Today is Tuesday" is a valid logical statement because it is either true or false, but "Mozart is the greatest composer" can not be assigned a truth value. It is an opinion.

For a given statement p, we define its *negation* to be the statement with the opposite truth value. We denote the negation of p by $\neg p$. Forming the negation of a simple statement is often very easy. For example, for the statement

p: *Today is Tuesday,*

the negation is

$\neg p$: *Today is not Tuesday.*

As another example, let $n \in \mathbb{N}$ and consider

q: $n^3 > 0.$

Its negation is

$$\neg q\colon n^3 \leq 0.$$

3.1.2 Combining statements

Given statements p and q, we can combine them to form the new statements $p \wedge q$ (*p and q*) and $p \vee q$ (*p or q*). The former is true only if both p is true and q is true. The latter is true if p is true, if q is true, or if both are true. Because such verbal descriptions can be cumbersome and confusing, we often use a *truth table* instead. Let's look first at the truth table for $p \wedge q$. The first two columns give all possible combinations of truth values for p and q. The final column for $p \wedge q$ indicates the truth value of $p \wedge q$ given the truth values for p and q in that row.

p	q	$p \wedge q$
T	T	T
T	F	F
F	T	F
F	F	F

Next, we show the truth table for $p \vee q$.

p	q	$p \vee q$
T	T	T
T	F	T
F	T	T
F	F	F

Remark 2. In English, the word *or* can be used in two different ways–as an *exclusive or* or as an *inclusive or*. To understand the difference, consider the following statements:

All dinners are served with soup or salad.
Everyone in this class has taken either Linear Algebra or Calculus III.

You probably understand the first sentence to mean that you may have soup with your dinner or you may have salad with your dinner, but you may not have both. On the other hand, you would not consider the second sentence to be false if there were a student who has taken both courses. Whether the "or" in an English sentence is to be interpreted as an exclusive or inclusive or depends on the context, and this can be a source of confusion. To avoid this confusion in mathematics we agree that we will use the inclusive or.

Example 3.1. We make a truth table for the statement $p \wedge \neg q$. The truth table only needs to have three columns–one for p, one for q, and one showing how the truth value of $p \wedge \neg q$ depends on the truth value of p and q. We find it useful, however, to include an intermediate column for $\neg q$.

p	q	$\neg q$	$p \wedge \neg q$
T	T	F	F
T	F	T	T
F	T	F	F
F	F	T	F

Exercise 3.1. Make a truth table for each statement:

(a) $\neg(p \wedge q)$

(b) $\neg(p \vee q)$

(c) $\neg p \wedge \neg q$

(d) $\neg p \vee \neg q$.

Look at the final columns. What do you observe?

3.1.3 Implications

Next we discuss one of the most important logical relations between two statements, namely the *implication* $p \implies q$, read "p implies q" or "if p, then q." For example, for the statements

p: Today is Monday
q: The cafeteria is serving pizza for lunch,

the statement $p \implies q$ is

If today is Monday, then the cafeteria is serving pizza for lunch.

The truth table for $p \implies q$ follows.

p	q	$p \implies q$
T	T	T
T	F	F
F	T	T
F	F	T

In the implication $p \implies q$, we call p the *hypothesis* and q the *conclusion.* Note that $p \implies q$ is only false when the hypothesis p is true but the conclusion q is false.

In terms of our example, $p \implies q$ is true if it is indeed Monday and the cafeteria is serving pizza. If it is Monday, but the cafeteria is not serving pizza, $p \implies q$ is false. This much seems to be common sense. But what truth value should we attach to $p \implies q$ if the hypothesis p is false, that is, if it is not Monday? We agree that if the hypothesis is false, we will interpret $p \implies q$ as true.

Given an implication $p \implies q$, we can form three related implications.

Definition 3.1. Let p and q be statements. The **converse** of $p \implies q$ is the implication $q \implies p$, the **inverse** of $p \implies q$ is $\neg p \implies \neg q$, and the **contrapositive** of $p \implies q$ is $\neg q \implies \neg p$.

For example, the converse of "If it is Monday, then the cafeteria is serving pizza for lunch" is

If the cafeteria is serving pizza for lunch, then it is Monday.

Two logical statements are called *equivalent* if one is true precisely when the other is true, i.e., if their truth tables are identical. It is intuitively clear from this example that a statement and its converse are not equivalent; the fact that if it is Monday the cafeteria is serving pizza does not mean that the cafeteria *only* serves pizza on Monday. To see in general that a statement and its converse are not logically equivalent, let's look at the truth tables for both.

p	q	$p \implies q$	$q \implies p$
T	T	T	T
T	F	F	T
F	T	T	F
F	F	T	T

For some choices of statements p and q, both of the implications $p \implies q$ and $q \implies p$ are true. In this case, we write $p \iff q$, read "p if and only if q."

For our example, the inverse of "If it is Monday, then the cafeteria is serving pizza" is

If it is not Monday, then the cafeteria is not serving pizza.

Again, our intuition tells us that the inverse of an implication is not logically equivalent to the implication. The truth tables confirm this intuition. We add some intermediate columns for $\neg p$ and $\neg q$ to make filling in the final column easier.

p	q	$\neg p$	$\neg q$	$p \implies q$	$\neg p \implies \neg q$
T	T	F	F	T	T
T	F	F	T	F	T
F	T	T	F	T	F
F	F	T	T	T	T

The contrapositive of the implication "If it is Monday, then the cafeteria is serving pizza" is

If the cafeteria is not serving pizza, then it is not Monday.

The contrapositive is indeed logically equivalent to the original statement, as the truth tables show.

p	q	$\neg p$	$\neg q$	$p \implies q$	$\neg q \implies \neg p$
T	T	F	F	T	T
T	F	F	T	F	F
F	T	T	F	T	T
F	F	T	T	T	T

Exercise 3.2. For each of the implications below, identify the hypothesis p and the conclusion q. Then form the converse, inverse, and contrapositive of each implication.

(a) Let n be an integer. If $n \leq 1$, then $n^2 \leq 1$.

(b) Let m and n be integers. If $m > 0$ and $n > 0$, then $mn > 0$.

(c) If a person was born in the United States, then that person is a United States citizen.

Exercise 3.3. Show that $\neg(p \implies q)$ is equivalent to $p \wedge \neg q$ by making the truth tables.

3.2 Quantified Statements and Their Negations

Universally quantified statements are statements about *all* elements of a set, such as

For any triangle, the angle measures sum to $180°$.
For every integer n, $n^2 > 0$.
All leopards have spots.

Generally, universally quantified statements can be written in the form

For all $x \in S$, p,

where p is a statement. Sometimes, in place of "for all" we use the symbol $\forall$ and write

$$\forall x \in S,\ p.$$

We will not always use precisely this wording. Phrases like "for all," "for every," "for any," and "for each" are synonymous. A universally quantified statement can even have the form of the last statement, "All leopards have spots," which could be (pedantically) rewritten as

Let S *be the set of leopards. For all* $x \in S$, x *has spots.*

Existentially quantified statements are statements about the existence of an element of a set with a certain property, such as

There exists a rational number r *such that* $r^2 = 144$.
There exists a continuous function $f\colon (0,1) \to \mathbb{R}$ *that is unbounded.*
There exists a leopard with no spots.

Sometimes, in place of "there exists" we use the symbol $\exists$. Generally, an existentially quantified statement has the form

$$\exists x \in S \text{ such that } p,$$

where p is a statement.

Example 3.2. Some statements are very complicated, involving several quantifiers, such as

For every $a \in \mathbb{R}$ *with* $a > 0$*, there exists a real number* x *such that* $x^2 = a$,

or

For every $\varepsilon > 0$*, there exists* $N \in \mathbb{N}$ *such that, for all* $n \geq N$, $2^{-n} < \varepsilon$.

For every mathematical statement, it is either true or its negation is true. Thus when you are trying to do mathematics, you are often simultaneously considering a statement and its negation, trying to determine which one is true. It is therefore very important to be able to negate even very complicated mathematical statements.

How do we form the negation of a universally quantified statement? Consider, for example, the statement "All leopards have spots." In order for this statement to be false (and its negation true), it need not be the case that no leopard has spots; the statement will be false if there is a single leopard without spots. Thus the negation is

There exists a leopard with no spots.

Similarly, the negation of

For every integer n, $n^2 > 0$

is

There exists an integer n *such that* $n^2 \leq 0$.

In general, the negation of the universally quantified statement

$$\forall x \in S,\ p$$

is the existentially quantified statement

$$\exists x \in S \textit{ such that } \neg p.$$

On the other hand, the negation of the existentially quantified statement

$$\exists x \in S \textit{ such that } p$$

is the universally quantified statement

$$\forall x \in S,\ \neg p.$$

Of course we can negate a complicated statement with nested quantifiers.

Example 3.3. Let us negate the second statement in Example 3.2, namely

For every $\varepsilon > 0$*, there exists* $N \in \mathbb{N}$ *such that, for all* $n \geq N$, $2^{-n} < \varepsilon$.

Note that this statement is of the form

For every $\varepsilon > 0$, p,

where p is the statement

p: *There exists* $N \in \mathbb{N}$ *such that, for all* $n \geq N$, $2^{-n} < \varepsilon$.

Thus the negation is

There exists $\varepsilon > 0$ *such that* $\neg p$.

Our task is now to negate p, which is itself a quantified statement of the form

p: *There exists* $N \in \mathbb{N}$ *such that* q,

where q is the statement

q: *For all* $n \geq N$, $2^{-n} < \varepsilon$.

Thus the negation of p is

$\neg p$: *For all* $N \in \mathbb{N}$, $\neg q$.

Finally, we need to negate q:

$\neg q$: *There exists* $n \geq N$ *such that* $2^{-n} \geq \varepsilon$.

Putting the pieces together gives the negation of the original statement:

There exists $\varepsilon > 0$ *such that, for all* $N \in \mathbb{N}$, *there exists* $n \geq N$ *such that* $2^{-n} \geq \varepsilon$.

Exercise 3.4. Negate each of the following statements.

(a) All students take calculus.

(b) There exists a natural number n for which $n > n^2$.

(c) There exists $M \in \mathbb{N}$ such that, for every $n \in \mathbb{N}$, $|a_n| \leq M$.

3.2.1 Writing implications as quantified statements

Consider the following statement about an integer n:

If n *is an integer, then* $n^2 > 0$.

This statement is written in the form of an implication $p \implies q$ where p is the statement "n is an integer" and q is the statement "$n^2 > 0$." We could also write $p \implies q$ as a universally quantified statement:

For all $n \in \mathbb{Z}$, $n^2 > 0$.

Rewriting the implication as a universally quantified statement may make it clearer to us what we must do to prove that it is false; we must prove its negation:

There exists $n \in \mathbb{Z}$ such that $n^2 \leq 0$.

Exercise 3.5. Write each implication as a universally quantified statement. Then negate the statement. Finally, determine whether each statement is true or false. If it is true, give a proof. If it is false, give a counterexample.

(a) If n is an even integer, n^2 is divisible by 4.

(b) If $x, y \in \mathbb{R}$, $x^3 + y^3 + 1 \geq 1$.

(c) Let $x \in \mathbb{R}$. If $|x - 2| < \frac{1}{100}$, $|x^2 - 4| < \frac{1}{20}$.

3.3 Proof Techniques

Now that we have a deeper understanding of the logic of mathematical statements and implications, we discuss some standard proof techniques. In each case, we give some examples and some basic exercises. Several of these techniques have already been used in this text.

3.3.1 Direct proof

The most obvious way to prove that p implies q is to assume that p is true and to show that q follows. Such a proof is called a *direct proof*. We have already given many direct proofs, but let's look at another elementary example.

Proposition 3.1. *Let $m, n \in \mathbb{Z}$. If m and n are odd, then $m + n$ is even.*

Proof. Because m and n are odd, there exist integers j and k such that $m = 2j + 1$ and $n = 2k + 1$. Then,

$$\begin{aligned} m + n &= (2j + 1) + (2k + 1) \\ &= 2j + 2k + 2 \\ &= 2(j + k + 1). \end{aligned}$$

Because $j + k + 1$ is an integer, $m + n$ is indeed divisible by 2. That is, $m + n$ is even. □

Exercise 3.6. Let $n \in \mathbb{Z}$. Give a direct proof of the proposition: If n is odd, n^2 is odd.

3.3.2 Proof by contradiction

Here's the idea: we want to prove statement p, but we can't find a direct proof. We thus suppose instead that p is false and show that this assumption implies some other statement we *know* to be false. We may then conclude that p is true. We have already seen this proof technique twice, in each direction of our proof of Proposition 2.3. We illustrate the technique again with a very simple example.

Proposition 3.2. *There is no smallest positive rational number.*

Proof. The proof is by contradiction. Thus suppose there is a smallest positive rational number r. Consider $a = \frac{1}{2}r$. Because the rational numbers are closed under multiplication, a is rational. Furthermore, because $0 < \frac{1}{2} < 1$ and $r > 0$, $0 < \frac{1}{2}r < r$, or, equivalently, $0 < a < r$. Thus we have found a positive rational number smaller than r. Because r was assumed to be the smallest positive rational number, we have a contradiction. We conclude that there is no smallest positive rational number. □

Exercise 3.7. Let p be prime and let $a = p! + 1$. Prove by contradiction that, for all integers $2 \leq k \leq p$, k does not divide a.

Let's use the last exercise to do another classic proof by contradiction.

Proposition 3.3. *The set of prime natural numbers has infinitely many elements.*

Proof. The proof is by contradiction. Thus suppose the set of primes has only finitely many elements. Then there is a largest prime p. Let $a = p! + 1$ and note that $a > p$. By Exercise 3.7, a is not divisible by any integer k with $2 \leq k \leq p$. Thus either a is itself prime or a is divisible by some prime larger than p. In either case, this contradicts the assumption that p is the largest prime. We may therefore conclude that the set of primes has infinitely many elements. □

3.3.3 Proof by contraposition

Recall that if p and q are statements, the implication $p \implies q$ is equivalent to its contrapositive, $\neg q \implies \neg p$. Thus we may always prove an implication by instead proving its contrapositive.

We illustrate with an example. Recall that two integers have the same *parity* if either both are even or both are odd.

Proposition 3.4. *If $m, n \in \mathbb{Z}$ and if $m + n$ is even, then m and n have the same parity.*

Proof. We prove the contrapositive; that is, we prove that if m and n do not have the same parity, then $m + n$ is not even. Because m and n do not have the same parity, one of them is even and the other is odd. We may assume without loss of generality that m is even and n is odd. Thus there exist integers j and k such that $m = 2j$ and $n = 2k + 1$. Then

$$m + n = 2j + (2k + 1) = 2(j + k) + 1.$$

Because $j + k$ is an integer, we see that $m + n$ is odd, as desired. □

Remark 3. We said, "We may assume without loss of generality that m is even and n is odd." Such an assumption is valid because we know that one of the integers is even and the other is odd. Were it the case that m is odd and n is even, we could simply rename our integers, calling the even one m and the odd one n. Such an assumption in a proof avoids having to separately treat two cases that are really identical except for the names of the objects.

Exercise 3.8. Let $n \in \mathbb{Z}$. Consider the proposition: If n^2 is even, then n is even. Prove this proposition by proving its contrapositive.

3.3.4 The art of the counterexample

Suppose you are presented with a statement that you believe to be false. How do you prove it? Because every statement must be true or false and because a statement and its negation can not have the same truth value, to prove that a statement is false, you must prove that its negation is true. Doing so may be a lot of work. Consider the statement "There exists a purple cow." This statement is existentially quantified. We could write it very formally as follows:

Let C be the set of all cows. Then there exists $c \in C$ such that c is purple.

Its negation is universally quantified:

For all $c \in C$, c is not purple.

Proving this statement would be hard, for it would require that we examine all cows.

Suppose, on the other hand, we are given a universally quantified statement that we believe to be false, such as "All roses are red." Formally,

Let R be the set of all roses. For all $r \in R$, r is red.

Its negation is

There exists $r \in R$ such that r is not red.

To prove the negation, all we need do is exhibit a rose that is not red, i.e., to provide a *counterexample* to the original statement. Of course, counterexamples may be hard to come by; even if "All clovers have three leaves" is false, it may be difficult to find a four-leafed clover.

Finding good counterexamples is a bit of an art. Consider the statement

For every $x, y \in \mathbb{R}$, $(x + y)^3 = x^3 + y^3$.

Our experience suggests to us that this statement is false. Let us analyze two possible arguments.
First Argument. The distributive law gives

$$(x+y)^3 = (x+y)(x+y)(x+y) = x^3 + 3x^2y + 3xy^2 + y^3.$$

Thus the original statement is true if and only if, for all $x, y \in \mathbb{R}$,

$$x^3 + 3x^2y + 3xy^2 + y^3 = x^3 + y^3.$$

This statement is, in turn, true if and only if, for all $x, y \in \mathbb{R}$,

$$3xy(x+y) = 0.$$

If $x, y \neq 0$, this last equality holds only if $x + y = 0$, i.e., only if $x = -y$. Therefore equality will not hold if, for example, $x = y = 1$. Thus the original statement does not hold for arbitrary $x, y \in \mathbb{R}$.
Second Argument. Consider $x = y = 1$. Then $(x+y)^3 = (1+1)^3 = 2^3 = 8$, but $x^3 + y^3 = 1^3 + 1^3 = 2$. Thus the original statement is false.

Both arguments are valid, but they differ greatly. The first may initially seem better because it begins by finding another expression equal to $(x+y)^3$. But it then leaves us with the new problem of determining whether $x^3+3x^2y+3xy^2+y^3$ does or does not equal x^3+y^3, or, equivalently, whether $3x^2y+3xy^2 = 0$ for all $x, y \in \mathbb{R}$. Our experience suggests that the latter statement is false, but, again, we need a proof. In this case, we are able to obtain such a proof. But imagine how much harder this approach would be were we to try to use it to prove that the statement

$$(x+y)^7 = x^7 + y^7$$

is false. At some point we would be forced to do what we do right away in the second argument, namely find specific values of x and y for which some statement fails. Our second argument above is more satisfying; the two expressions are unequal for "most" values of x and y, and so it is a simple matter to find values that are easy to substitute to illustrate that the expressions on the left and right are unequal.

Exercise 3.9. Each of the following statements is false. Provide a simple counterexample to each.

(a) For all $x, y \in \mathbb{R}$, $(x+y)^2 = x^2 + y^2$.

(b) For all $x, y > 0$, $\frac{1}{x+y} = \frac{1}{x} + \frac{1}{y}$.

(c) For all $n, m \in \mathbb{N}$, $\sqrt{n^2 + m^2} = n + m$.

3.4 Problems

1. Consider the following two statements:

For every $x \in \mathbb{R}$, there exists $y \in \mathbb{R}$ such that $y^2 - x^2 = 1$.
There exists $y \in \mathbb{R}$ such that, for every $x \in \mathbb{R}$, $y^2 - x^2 = 1$.

(a) Explain carefully the difference between the two statements. Are both true?

(b) Negate the above statements. Is either of the negations true?

2. Let $m, n \in \mathbb{Z}$. Determine whether each of the following statements is true or false. Then form the converse of each and determine whether it is true or false. In each case, if the statement is true, give a proof, and if it is false, give a counterexample.

   (a) If mn is even, then m and n are even.

   (b) If n and m are both divisible by 3, then $m + n$ is divisible by 3.

   (c) If $m < n$, $m^2 < n^2$.

3. Consider the following statements about $n \in \mathbb{N}$.

   p: n is divisible by 3.
   q: The sum of the digits in the base 10 *representation of n is divisible by* 3.

   Does p imply q? Does q imply p?

4. A sequence $\{a_n\}$ is bounded if *there exists M such that, for all $n \in \mathbb{N}$, $|a_n| \leq M$*. Negate the italicized statement to obtain the definition of a sequence $\{a_n\}$ that is *not bounded.* Then prove that the sequence with n-th term given by

$$a_n = \begin{cases} 0 & \text{if } n \text{ is even} \\ \frac{1}{2}n & \text{if } n \text{ is odd} \end{cases}$$

   is not bounded.

5. Let $\{a_n\}$ and $\{b_n\}$ be sequences of real numbers. Determine whether each of the following statements is true or false. As usual, if the statement is true, prove it; if it is false, give a counterexample.

   (a) If $\{a_n\}$ and $\{b_n\}$ are bounded, then $\{a_n b_n\}$ is bounded.

   (b) If $\{a_n b_n\}$ is bounded, then $\{a_n\}$ and $\{b_n\}$ are bounded.

   (c) If $\{a_n + b_n\}$ is bounded, then $\{a_n\}$ and $\{b_n\}$ are bounded.

Programming Project. A hallway in a high school has 100 lockers. Each locker has two possible states: open and closed. Initially, all lockers are closed. One hundred students enter the hallway one at a time. The first changes the state of every locker (thus opens every locker). The second changes the state of every second locker. The third changes the state of every third locker, and, in general, the n-th student changes the state of every n-th locker.

1. By exploring the problem by hand, make a conjecture about which lockers will be open after all 100 students have gone through the hallway.

2. Write a program that produces as output the numbers of all lockers that are open after all 100 students have gone through the hallway.

3. Now suppose the hallway has N lockers, initially closed, and that N students pass through the hallway as described above. Make a conjecture about which lockers will be open after all N students have gone through the hallway. Prove your conjecture.

Chapter 4

Sets, Relations, and Functions

In Chapter 1, we talked briefly about sets and functions, for the most part just establishing basic notation and terminology. In this chapter we discuss these notions in more detail, establishing the basic results and developing the basic proof techniques.

4.1 Sets

In Chapter 1, we defined the union and intersection of two sets. These definitions can be extended to arbitrary collections of sets. The union of a collection of sets consists of those objects that are elements of at least one set in the collection, whereas the intersection of a collection of sets consists of those objects that are elements of all the sets in the collection. We make the formal definition and establish our notation.

Definition 4.1. Let $\{S_\alpha : \alpha \in \mathcal{A}\}$ be a collection of sets. Then their **union** is

$$\bigcup_{\alpha \in \mathcal{A}} S_\alpha = \{x : \exists \alpha \in \mathcal{A} \text{ such that } x \in S_\alpha\}$$

and their **intersection** is

$$\bigcap_{\alpha \in \mathcal{A}} S_\alpha = \{x : \forall \alpha \in \mathcal{A}, x \in S_\alpha\}.$$

Given two sets A and B, perhaps the most basic question to consider is whether A equals B. The most obvious way to show that $A = B$ is to show that (i) every element of A is an element of B and (ii) every element of B is an element of A. That is, to show $A = B$, we prove the two containments $A \subseteq B$ and $B \subseteq A$.

Example 4.1. We claim that

$$\bigcup_{n=1}^{\infty}\left[-1+\frac{1}{n}, 1-\frac{1}{n}\right]=(-1,1). \tag{4.1}$$

We prove two containments. Let A denote the union on the left-hand side of (4.1). If $x \in A$, then there exists $N \in \mathbb{N}$ such that $x \in \left[-1+\frac{1}{N}, 1-\frac{1}{N}\right]$. Then

$$-1<-1+\frac{1}{N} \leq x \leq 1-\frac{1}{N}<1,$$

and $x \in(-1,1)$. Thus $A \subseteq(-1,1)$.

For the reverse containment, suppose $x \in(-1,1)$. Then $1-|x|>0$ and so, by the Archimedean principle, there exists $N \in \mathbb{N}$ such that $1-|x| \geq \frac{1}{N}>0$. This inequality is equivalent to

$$1-\frac{1}{N} \geq|x|,$$

which is in turn equivalent to

$$-1+\frac{1}{N} \leq x \leq 1-\frac{1}{N}.$$

Thus $x \in\left[-1+\frac{1}{N}, 1-\frac{1}{N}\right]$ and $x \in A$. That is, we have shown that $(-1,1) \subseteq A$. Because $A \subseteq(-1,1)$ and $(-1,1) \subseteq A$, $A=(-1,1)$.

Exercise 4.1. Let

$$B=\bigcap_{n=1}^{\infty}\left(-1-\frac{1}{n}, 1+\frac{1}{n}\right).$$

Show that $B=[-1,1]$.

Our next proposition establishes a result that is used often.

Proposition 4.1 (De Morgan's Laws). *Let $\{S_\alpha : \alpha \in \mathcal{A}\}$ be a collection of sets.*

(a) $$\left(\bigcup_{\alpha \in \mathcal{A}} S_\alpha\right)^c=\bigcap_{\alpha \in \mathcal{A}} S_\alpha^c.$$

(b) $$\left(\bigcap_{\alpha \in \mathcal{A}} S_\alpha\right)^c=\bigcup_{\alpha \in \mathcal{A}} S_\alpha^c.$$

Proof. We prove part (a) and leave part (b) as the next exercise.

Our set identity essentially follows from negating the defining property for $\bigcup_{\alpha \in \mathcal{A}} S_\alpha$; indeed, $x \notin \bigcup_{\alpha \in \mathcal{A}} S_\alpha$ if and only if, for all $\alpha \in \mathcal{A}$, $x \notin S_\alpha$, i.e., if and only if for all $\alpha \in \mathcal{A}$, $x \in S_\alpha^c$. Thus $x \in \left(\bigcup_{\alpha \in \mathcal{A}} S_\alpha\right)^c$ if and only if $x \in \bigcap_{\alpha \in \mathcal{A}} S_\alpha^c$. □

Exercise 4.2. Prove part (b) of Proposition 4.1.

We discuss another way to combine sets.

Definition 4.2. Let A and B be sets. Let

$$A \times B = \{(a, b) : a \in A,\ b \in B\}.$$

We call $A \times B$ the **Cartesian product** of A and B.

In other words, $A \times B$ is the set of all *ordered pairs* in which the first element is in A and the second element is in B.

Remark 4 (open interval versus ordered pair). Frustratingly, we now have two possible meanings for (a, b); it may be an open interval in $\mathbb{R}$ or it may be an element of $A \times B$. It should be clear from the context which is meant because intervals and ordered pairs are two different kinds of objects. Intervals are sets, whereas ordered pairs are elements of sets. Thus if we write $(1, 3) \cap (2, 4)$, $(1, 3)$ and $(2, 4)$ must be sets, and so in this case they are open intervals. On the other hand, if we write $(1, 3) \in \mathbb{N} \times \mathbb{N}$, $(1, 3)$ is an ordered pair of natural numbers.

You are already very familiar with one set that is a Cartesian product, namely the Cartesian plane

$$\mathbb{R}^2 = \mathbb{R} \times \mathbb{R} = \{(x, y) : x, y \in \mathbb{R}\}.$$

Let's consider a less standard example.

Example 4.2. Let $A = \{a, b\}$ and $B = \{1, 2, 3\}$. Then

$$A \times B = \{(a, 1), (a, 2), (a, 3), (b, 1), (b, 2), (b, 3)\}.$$

Exercise 4.3. (a) For the sets A and B in Example 4.2, find $B \times A$.

(b) Under what conditions on A and B will $A \times B = B \times A$? (Prove your assertion.)

4.2 Relations

4.2.1 The definition

Definition 4.3. Let A and B be non-empty sets. A **relation** between A and B is a subset R of $A \times B$. If $A = B$, we call R a **relation on** A.

Intuitively, a relation is a rule for deciding whether a pair (a, b) (with $a \in A$ and $b \in B$) is or is not "related." Some examples will help clarify.

Example 4.3. Let $A = \{\text{California, Oregon, Washington, Texas}\}$ and let $B = \{\text{Los Angeles, Portland, Seattle, Olympia, Chicago, San Diego, Madison}\}$. We define a relation R between A and B by declaring that (a, b) is in R if b is a city in A. Thus the elements of R are: (California, Los Angeles), (California, San Diego), (Oregon, Portland), (Washington, Seattle), and (Washington, Olympia). Note that it need not be the case that every element of A appears as the first entry of some element of R or that every element of B appears as the second entry of some element of R.

When R is a relation between A and B, we sometimes write aRb instead of $(a, b) \in R$.

Exercise 4.4. Let $A = B = \mathbb{N}$. Define a relation R on $\mathbb{N}$ by $(a, b) \in R$ if a divides b.

(a) For which a is $(a, 12)$ in R?

(b) For which b is $(1, b)$ in R? For which a is $(a, 1)$ in R?

4.2.2 Order relations

Let's discuss a kind of relation that is very familiar to you.

Definition 4.4. Let A be a non-empty set. Then $<$ is an **order relation** on A if the following conditions hold:

(O1) (Comparability) For all $a, b \in A$ with $a \neq b$, either $a < b$ or $b < a$.

(O2) (Non-reflexivity) There is no $a \in A$ for which $a < a$.

(O3) (Transitivity) For all $a, b, c \in A$, if $a < b$ and $b < c$, then $a < c$.

As with most definitions in abstract mathematics, this one is motivated by something very familiar; the canonical example of an order relation is the usual relation $<$ defined on $\mathbb{R}$ or on one of its subsets. We consider several other examples and non-examples.

Example 4.4. Let A be the set of all finite strings of letters. If s_1 and s_2 are in A, we say $s_1 < s_2$ if s_1 comes before s_2 in alphabetical order. Alphabetical order is an order relation; any two distinct letter strings are comparable, it does not make sense to say that a string comes before itself alphabetically, and if s_1 comes before s_2 and s_2 comes before s_3, then s_1 comes before s_3.

Example 4.5. Let S be the set of all subsets of $\{1, 2, 3\}$ and consider the relation $\subset$ on S. This relation is non-reflexive, for a subset A of S can not be a proper subset of itself. The relation is also transitive, for if A, B, and C are subsets of S with $A \subset B$ and $B \subset C$, then $A \subset C$. However, the relation does not satisfy the comparability property. Consider $A = \{1, 2\}$ and $B = \{2, 3\}$. Neither $A \subset B$ nor $B \subset A$ is true. We conclude that $\subset$ is not an order relation on S.

Exercise 4.5. Let A be the set of all living people. Define a relation $<$ on A by declaring that $a < b$ if b is older than a. Is $<$ an order relation on A?

4.2.3 Equivalence relations

Next we discuss an incredibly important kind of relation that will appear throughout this book.

Definition 4.5. Let A be a set. A relation $\sim$ on A is an **equivalence relation** if the following conditions hold:

(E1) (reflexivity) For all $a \in A$, $a \sim a$.

(E2) (symmetry) For all $a, b \in A$, if $a \sim b$, then $b \sim a$.

(E3) (transitivity) For all $a, b, c \in A$, if $a \sim b$ and $b \sim c$, then $a \sim c$.

We illustrate the definition with one of the most commonly considered equivalence relations. We will study equivalence relations like this one in detail in Chapter 6.

Example 4.6. We define a relation $\sim$ on $\mathbb{Z}$ by the rule $n \sim m$ if 3 divides $n - m$. We claim that $\sim$ is an equivalence relation on $\mathbb{Z}$. To prove this claim, we must verify that $\sim$ satisfies properties (E1)–(E3) in Definition 4.5.

First, observe that, for all $n \in \mathbb{Z}$, $n - n = 0 = 3(0)$. Thus $n \sim n$, proving reflexivity of the relation.

Next, suppose that $n, m \in \mathbb{Z}$ and that $n \sim m$. Thus there exists $k \in \mathbb{Z}$ such that $n - m = 3k$. Then $m - n = -(n - m) = -3k = 3(-k)$. Thus $m \sim n$, proving symmetry.

Finally, suppose $n, m, p \in \mathbb{Z}$, with $n \sim m$ and $m \sim p$. Thus there exist integers j and k such that $n - m = 3j$ and $m - p = 3k$. Then

$$n - p = n - m + m - p = 3j + 3k = 3(j + k).$$

Thus $n \sim p$, proving transitivity.

Example 4.7. Let S be the set of all students at this university. We define a relation R on S by

$$R = \{(a, b) \in S \times S : a \text{ and } b \text{ are in a class together}\}.$$

We claim that R is not an equivalence relation. In particular, R is not transitive, for it could be the case that Amy and Bill are in Linear Algebra together and Bill and Celeste are in Organic Chemistry together, but that Amy and Celeste do not share a class.

Exercise 4.6. Determine whether the relation from Exercise 4.4 is an equivalence relation.

Exercise 4.7. Let P be the set of all people, living or deceased. Give an example of a relation on P that is transitive but not symmetric.

When we have an equivalence relation $\sim$ on a set A, then for each $a \in A$, we can form the set of all elements of A equivalent to it.

Definition 4.6. Let A be a set with an equivalence relation $\sim$. Then the **equivalence class containing** a is the set

$$\bar{a} = \{x \in A : x \sim a\}.$$

Example 4.8. Consider the equivalence relation in Example 4.6. Then

$$\begin{aligned}\bar{1} &= \{n \in \mathbb{Z} : n \sim 1\} \\ &= \{n \in \mathbb{Z} : n - 1 \text{ is a multiple of } 3\} \\ &= \{n \in \mathbb{Z} : n = 3k + 1 \text{for some } k \in \mathbb{Z}\} \\ &= \{\ldots, -5, -2, 1, 4, 7, \ldots\}.\end{aligned}$$

Exercise 4.8. Consider the equivalence relation in Example 4.6.

(a) Describe the equivalence classes $\bar{0}$ and $\bar{2}$.

(b) Describe the equivalence class $\bar{7}$.

(c) How many distinct equivalence classes are there for this equivalence relation?

As the previous exercise illustrates, an equivalence relation *partitions* a set A into disjoint equivalence classes; everything in A is obviously in at least one equivalence class, and if two equivalence classes $\bar{a}$ and $\bar{b}$ share an element, then necessarily $\bar{a} = \bar{b}$. We prove this latter assertion as a proposition.

Proposition 4.2. *Let A be a set with an equivalence relation $\sim$ and let $\bar{a}$ and $\bar{b}$ be the equivalence classes determined by elements a and b of A. If $\bar{a} \cap \bar{b} \neq \emptyset$, then $\bar{a} = \bar{b}$.*

Proof. We prove the set identity by proving two containments.

Suppose $x \in \bar{a}$. Then by definition, $x \sim a$. Because $\bar{a} \cap \bar{b}$ is not empty, there exists $c \in \bar{a} \cap \bar{b}$. Because $c \in \bar{a}$, by definition $c \sim a$. By symmetry, $a \sim c$, and then by transitivity $x \sim c$. Because c is also in $\bar{b}$, $c \sim b$, and so by transitivity again, $x \sim b$, i.e., $x \in \bar{b}$. We have now shown that $\bar{a} \subseteq \bar{b}$.

The proof that $\bar{b} \subseteq \bar{a}$ is the same and is left to the reader. □

Exercise 4.9. Complete the proof of the above proposition by showing that $\bar{b} \subseteq \bar{a}$.

4.3 Functions

4.3.1 Images and pre-images

Let A and B be non-empty sets and let $f\colon A \to B$. We have already defined $f(A)$, the image of A under f. We can extend this definition to arbitrary subsets of the domain. Thus if $C \subseteq A$, we define $f(C)$, the *image* of C under f, by

$$f(C) = \{b \in B : \exists c \in C \text{ such that } f(c) = b\}.$$

For a subset D of B, we can define $f^{-1}(D)$, its *pre-image* under f, by

$$f^{-1}(D) = \{a \in A : f(a) \in D\}.$$

Example 4.9. Consider $f\colon \mathbb{R} \to \mathbb{R}$ defined by $f(x) = x^2$. Let $C = \{x \in \mathbb{R} : 0 < x < 2\}$. In interval notation, $C = (0, 2)$. Let $D = (0, 4)$. Then

$$f(C) = (0, 4)$$

but

$$f^{-1}(D) = (-2, 0) \cup (0, 2).$$

In particular, note that $f^{-1}(f(C)) \neq C$!

Exercise 4.10. Consider $f\colon \mathbb{R} \to \mathbb{R}$ defined by $f(x) = x^3 - 1$. Find $f([0, 2])$, $f^{-1}(\{26\})$, and $f(f^{-1}([-1, 0]))$.

Exercise 4.11. (a) Let $f\colon A \to B$ and let $D \subseteq B$. Show that $f(f^{-1}(D)) \subseteq D$.

(b) Is it true in general that $D = f(f^{-1}(D))$? (As always, give a proof or a counterexample.)

4.3.2 Injections, surjections, and bijections

Let A and B be non-empty sets. We define three special classes of functions.

Definition 4.7. $f\colon A \to B$ is an **injection** if, for every $a_1, a_2 \in A$ with $a_1 \neq a_2$, $f(a_1) \neq f(a_2)$. This statement is equivalent to the statement that, if $f(a_1) = f(a_2)$, then $a_1 = a_2$.

Definition 4.8. $f\colon A \to B$ is a **surjection** if, for every $b \in B$, there exists $a \in A$ such that $f(a) = b$.

Definition 4.9. $f\colon A \to B$ is a **bijection** if it is both an injection and a surjection.

If a function f is an injection, we say that f is *injective* or, equivalently, *one-to-one.* If f is a surjection, we say that f is *surjective* or, equivalently, *onto.* In simple language, f is injective if distinct elements of the domain are mapped to distinct elements of the range, and f is surjective if every element in the range is actually in the image of the function.

Exercise 4.12. Let $f\colon A \to B$.

(a) By negating the defining statement in the definition above, say precisely what it means for f to fail to be an injection.

(b) By negating the defining statement in the definition above, say precisely what it means for f to fail to be a surjection.

Example 4.10. Consider $f\colon \mathbb{R} \setminus \{1\} \to \mathbb{R}$ defined by $f(x) = \frac{1}{x-1}$. We claim that f is injective. Indeed, suppose $x_1, x_2 \in \mathbb{R} \setminus \{1\}$ and that $f(x_1) = f(x_2)$, that is,

$$\frac{1}{x_1 - 1} = \frac{1}{x_2 - 1}.$$

Because $x_1 - 1 \neq 0$ and $x_2 - 1 \neq 0$, we may multiply both sides of the equation by $(x_1 - 1)(x_2 - 1)$ to obtain the equivalent equation

$$x_2 - 1 = x_1 - 1.$$

Adding 1 to both sides yields $x_1 = x_2$, as desired.

This function is not surjective, for there is no $x \in \mathbb{R} \setminus \{1\}$ for which $f(x) = \frac{1}{x-1} = 0$.

Whether or not a function is injective or surjective depends on more than just the rule or formula that defines it; it depends on the rule *as well as on the domain and range.* The next examples illustrate this point.

Example 4.11. Consider $f \colon \mathbb{R} \to \mathbb{R}$ defined by $f(x) = (x-1)^2$. Then f is not injective, for if we take $x_1 = 2$ and $x_2 = 0$, then $f(x_1) = f(2) = (2-1)^2 = 1$ and $f(x_2) = f(0) = (0-1)^2 = 1$.

Now consider $g \colon (1, \infty) \to \mathbb{R}$ defined by $g(x) = (x-1)^2$. g is not the same function as f because it has a different domain. g *is* injective. Indeed, if $x_1, x_2 \in (1, \infty)$ and $g(x_1) = g(x_2)$, then $(x_1 - 1)^2 = (x_2 - 1)^2$. The fact that both $x_1 - 1$ and $x_2 - 1$ are non-negative implies that $x_1 - 1 = +\sqrt{(x_2-1)^2} = x_2 - 1$, and so $x_1 = x_2$.

Example 4.12. Consider $f \colon \mathbb{N} \to \mathbb{N}$ defined by $f(n) = n + 1$. Then f is injective, for if $f(n_1) = f(n_2)$, then $n_1 + 1 = n_2 + 1$, and it follows that $n_1 = n_2$. On the other hand, f is not surjective because $1 \in \mathbb{N}$ but there exists no natural number n for which $f(n) = n + 1 = 1$.

Next consider $g \colon \mathbb{N} \to \mathbb{N} \setminus \{1\}$ defined by $g(n) = n+1$. This function g is *not* the same as the function f from the last paragraph because f and g have different ranges. Because f and g are given by the same formula, the argument given to show that f is injective shows that g is injective. Our function g, however, is surjective. Indeed, if $m \in \mathbb{N} \setminus \{1\}$, then $m - 1 \in \mathbb{N}$ and $g(m-1) = m$.

Exercise 4.13. Give an example of a function $g \colon \mathbb{N} \to \mathbb{N}$ that is surjective but not injective.

Exercise 4.14. Clearly $\iota \colon \mathbb{N} \to \mathbb{N}$ defined by $\iota(n) = n$ is a bijection. This function is called the **identity function.** Give another example of a bijection from $\mathbb{N}$ to $\mathbb{N}$.

4.3.3 Compositions of functions

We now discuss the composition of two functions. This concept is no doubt familiar to you from calculus.

Definition 4.10. Let A, B, and C be non-empty sets and let $f \colon A \to B$ and $g \colon B \to C$. The **composition** of g with f is the function $g \circ f \colon A \to C$ defined by

$$(g \circ f)(a) = g(f(a)).$$

Composition of functions is an example of an operation that does not satisfy the commutative property. Take, for example, $f, g\colon \mathbb{R} \to \mathbb{R}$ given by $f(x) = x-1$ and $g(x) = 2x$. Then

$$(f \circ g)(x) = f(g(x)) = f(2x) = 2x - 1$$

but

$$(g \circ f)(x) = g(f(x)) = g(x-1) = 2(x-1) = 2x - 2.$$

Exercise 4.15. Give an example of a pair of distinct functions f and g for which $f \circ g = g \circ f$.

Our next proposition identifies some properties of functions that are preserved when we take compositions. We will use this proposition repeatedly in later chapters.

Proposition 4.3. *Let A, B, and C be non-empty sets and let $f\colon A \to B$ and $g\colon B \to C$.*

(a) If f and g are injective, then $g \circ f\colon A \to C$ is injective.

(b) If f and g are surjective, then $g \circ f\colon A \to C$ is surjective.

(c) If f and g are bijective, then $g \circ f\colon A \to C$ is bijective.

Proof. We prove part (a) and leave the proof of part (b) as an exercise. Note that part (c) follows immediately from parts (a) and (b), for if f and g are bijective, they are both injective and surjective, and so $g \circ f$ is both injective and surjective.

To prove part (a), suppose $a_1, a_2 \in A$ are such that $(g \circ f)(a_1) = (g \circ f)(a_2)$. By definition of composition,

$$g(f(a_1)) = g(f(a_2)).$$

Because g is injective, we may conclude that $f(a_1) = f(a_2)$. Furthermore, because f is also injective, we conclude that $a_1 = a_2$, as desired. □

Note how we approached this proof. We wanted to show that $g \circ f$ has a certain property, *so we looked to the definition of that property.* The definition says that a function is injective if, whenever two outputs are the same, the inputs must be the same. Thus we assumed we had a_1 and a_2 in the domain with $(g \circ f)(a_1)$ the same as $(g \circ f)(a_2)$ and we showed that this forced a_1 and a_2 to be the same. *Pay close attention to definitions. They tell you how to prove propositions.*

Exercise 4.16. Prove part (b) of Proposition 4.3.

4.3.4 Inverse functions

Suppose $f\colon A \to B$ is a bijection. Define a new function $g\colon B \to A$ by the rule

$$g(b) = a \iff f(a) = b.$$

Does this rule actually define a function from B to A? In order for g to be a function, we must check two things. First, it must be the case that g associates to every element of B some element of A. This condition will be satisfied if, for every $b \in B$ there exists $a \in A$ with $f(a) = b$. Because f is surjective, this condition is indeed satisfied. Second, it must be the case that g assigns to each b a *unique* value of a. When this condition is satisfied, we say that our function is *well-defined*. Our function g *is* well-defined by the injectivity of f. Indeed, if there exist $a_1, a_2 \in A$ with $g(b)$ associated with both a_1 and a_2, then $f(a_1) = b = f(a_2)$, and necessarily $a_1 = a_2$.

It follows from the definition of g that

$$\forall b \in B, f(g(b)) = b$$

and

$$\forall a \in A, g(f(a)) = a.$$

The above discussion justifies making the following definition.

Definition 4.11. Let $f\colon A \to B$ be a bijection. The **inverse function** is the function $f^{-1}\colon B \to A$ such that $f^{-1}(f(a)) = a$ for all $a \in A$ and $f(f^{-1}(b)) = b$ for all $b \in B$.

Exercise 4.17. Consider $f\colon \mathbb{R} \to \mathbb{R}$ defined by $f(x) = x^3 + 5$. f is a bijection.

(a) Find $f^{-1}(\{4\})$ and $f^{-1}(4)$. Explain what each notation means.

(b) Find $f^{-1}(y)$.

We end with a proposition whose proof we leave as an exercise.

Proposition 4.4. *Suppose $f\colon A \to B$ is a bijection. Then the inverse function $f^{-1}\colon B \to A$ is a bijection.*

Exercise 4.18. Prove Proposition 4.4.

4.4 Problems

1. Let A, B, and C be sets. Determine whether each of the following is true. As always, prove the true statements and give counterexamples for the false statements.

   (a) $A \cup (B \cap C) = (A \cup B) \cap (A \cup C)$.

   (b) $A \setminus B = B \setminus A$.

(c) $A \times (B \cup C) = (A \times B) \cup (A \times C)$.

2. Give a simple description of each of the following subsets of $\mathbb{R}$ (e.g., as a single interval or a set with a single point). Prove your assertions.

   (a) $\bigcup_{n=1}^{\infty} (-n, n)$.

   (b) $\bigcap_{n=1}^{\infty} (-n, n)$.

   (c) $\bigcap_{n=1}^{\infty} \left(-\frac{1}{n}, \frac{1}{n}\right)$.

3. Let P be the set of all people, living or deceased. Determine whether each relation R is an equivalence relation. If it is, explain why the three properties are satisfied. If it is not, state which properties fail.

   (a) $R = \{(a, b) \in P \times P : b \text{ is a descendant of } a\}$.

   (b) $R = \{(a, b) \in P \times P : a \text{ is a parent of } b\}$.

   (c) $R = \{(a, b) \in P \times P : a \text{ and } b \text{ have the same parents}\}$.

   (d) $R = \{(a, b) \in P \times P : a \text{ and } b \text{ are blood relatives}\}$.

4. Let $\mathbb{R}^2 = \{(x, y) : x, y \in \mathbb{R}\}$. Define a relation $<$ on $\mathbb{R}^2$ by $(x_1, y_1) < (x_2, y_2)$ if (i) $x_1 < x_2$ in the usual order relation on $\mathbb{R}$ or (ii) if $x_1 = x_2$ and $y_1 < y_2$ in the usual order relation on $\mathbb{R}$.

   (a) Determine whether each of the following is true or false.

      i. $(1, 0) < (2, -4)$.

      ii. $(1, 5) < (1, 4)$.

      iii. $(2, 2) < (2, 2)$.

   (b) Is $<$ an order relation on $\mathbb{R}^2$? Give a proof or show that one of the properties of an order relation is not satisfied.

5. Let $S = \{(a, b) : a \in \mathbb{Z}, b \in \mathbb{Z} \setminus \{0\}\}$. Define $\sim$ on S by the rule

$$(a, b) \sim (c, d) \iff ad = bc.$$

   (a) Show that $\sim$ is an equivalence relation on S.

   (b) Describe the equivalence class containing $(1, 2)$.

   (c) Describe the equivalence class containing $(-4, 10)$.

6. Let $S = \mathbb{R}^3 \setminus \{(0, 0, 0)\} = \{(x, y, z) : x, y, z \in \mathbb{R}\} \setminus \{(0, 0, 0)\}$. Define a relation $\sim$ on S by $(x, y, z) \sim (x', y', z')$ if and only if there exists a positive real number a such that $(x, y, z) = (ax', ay', az')$.

   (a) Show that $\sim$ is an equivalence relation on S.

(b) Describe the equivalence classes geometrically.

7. Let $f\colon A \to B$ and let $A_1, A_2 \subseteq A$ and $B_1, B_2 \subseteq B$. Determine whether each of the following statements is true or false.

   (a) $f(A_1 \cup A_2) = f(A_1) \cup f(A_2)$.

   (b) $f^{-1}(B_1 \cap B_2) = f^{-1}(B_1) \cap f^{-1}(B_2)$.

8. (a) Construct a bijection from $(-1, 1)$ to $\mathbb{R}$. (Hint: Look for a rational function. Begin by sketching a plausible graph.)

   (b) Construct a bijection from $\mathbb{R}$ to $(-1, 1)$.

9. Does there exist a bijection $f\colon \mathbb{Z} \to \mathbb{N}$? If so, give an example. If not, prove that no such bijection exists.

10. Consider $f\colon \mathbb{R} \to \mathbb{R}$ defined by

$$f(x) = ax^2 + bx + c$$

where a, b, and c are real constants.

   (a) For which, if any, values of a, b, and c is f injective?

   (b) For which, if any, values of a, b, and c is f surjective?

   (c) For which, if any, values of a, b, and c is f bijective?

11. Let $f\colon A \to B$ be a bijection and let $C \subseteq A$. Is it true that $f^{-1}(f(C)) = C$?

12. Let $f\colon \mathbb{R} \to \mathbb{R}$. Define $f^2 = f \circ f$, $f^3 = f \circ f \circ f$, and, in general, define $f^n = f \circ f \circ \ldots \circ f$, where the composition involves n copies of f. We call f^n the n-th iterate of f.

   (a) If $f(x) = x^2$, find and prove a formula for $f^n(x)$.

   (b) If $f(x) = 2x - 1$, find and prove formulas for $f^n(1)$, $f^n(2)$, and $f^n(4)$.

   (c) If $f(x) = m(x - b) + b$ for non-zero constants m and b, find and prove a formula for $f^n(x)$.

Programming Project. In this project, you will explore the iterates f^n of $f(x) = x^2 - 1$. (See Problem 12 for the definition of f^n.)

1. Find the terms of the sequence $\{x, f(x), f^2(x), f^3(x), \ldots\}$ by hand for $x = 0$ and $x = 1$.

2. Write a program that takes as input a value of x and plots the first 50 terms of the sequence $\{x, f(x), f^2(x), f^3(x), \ldots\}$.

3. Use your program to study the sequence $\{x, f(x), f^2(x), f^3(x), \ldots\}$ for $x = 1.1, 1.2, 1.3, 1.4, 1.5, 1.6, 1.7$. Describe your results.

4. By further exploration with different values of x, make a conjecture about the largest value of x for which $\{f^n(x)\}$ remains bounded.

5. A real number x is a *fixed point* of f if $f(x) = x$. Find all fixed points of $f(x) = x^2 - 1$. What can you say about $\{x, f(x), f^2(x), f^3(x), \ldots\}$ if x is a fixed point? Modify the conjecture you made above if necessary.

Chapter 5

Elementary Discrete Mathematics

Discrete means "individually separated or distinct." When used in the context of mathematics, it is the opposite of *continuous*. Thus whereas continuous mathematics is the study of the real line (the continuum), continuous functions on the line, the calculus of such functions, and the generalizations of these notions, discrete mathematics is the study of finite sets, sequences, recurrence relations, and other related notions. (It is not to be confused with *discreet mathematics*, which would be math done in hushed voices.) This chapter introduces some of the major topics in discrete mathematics, including elementary combinatorics, recurrence relations, and the analysis of algorithms. Such topics can be explored rather easily by hand or with simple programs, and mathematical induction is a commonly-used proof technique.

5.1 Basic Principles of Combinatorics

Combinatorics is the branch of mathematics concerned with counting finite sets. Counting may sound like a simple task, but in practice it can be rather difficult.

5.1.1 The Addition and Multiplication Principles

Our first basic principle of counting is the *addition principle*. This principle is so simple and intuitive that it barely needs to be stated.

Addition Principle. Suppose A and B are disjoint finite sets and that A has n elements and B has m elements. Then their union $A \cup B$ has $n+m$ elements.

This principle can be generalized. We first make two definitions.

Definition 5.1. If A is a set, a **partition** of A is a collection $\{A_\alpha : \alpha \in A\}$ of subsets of A such that

(i) for all $\alpha, \beta \in \mathcal{A}$ with $\alpha \neq \beta$, $A_\alpha \cap A_\beta = \emptyset$, and

(ii) $A = \bigcup_{\alpha \in \mathcal{A}} A_\alpha$.

Definition 5.2. For a finite set S, we denote by $|S|$ the number of elements of S. We call $|S|$ the **cardinality** of S.

We may now use the addition principle for two sets and induction to prove a generalization.

Proposition 5.1 (General Addition Principle). *Let $\{A_i : i = 1, 2, \ldots, n\}$ be a partition of a finite set A, and suppose $|A_i| = m_i$. Then*

$$|A| = \sum_{i=1}^{n} |A_i| = \sum_{i=1}^{n} m_i.$$

Proof. The proof is by induction on n. When $n = 1$, the result is obvious because our partition of A consists of only a single set, which is necessarily equal to A by property (ii) in the definition of a partition.

Suppose, then, that the result holds for some $k \geq 1$ and consider $k + 1$. Thus suppose we have a partition $\{A_i : i = 1, \ldots, k, k + 1\}$ of a set A. Let $B = \cup_{i=1}^{k} A_i$. Then $\{A_i : i = 1, \ldots, k\}$ is a partition of B, and $\{B, A_{k+1}\}$ is a partition of A. By the addition principle for two sets,

$$|A| = |B| + |A_{k+1}| = |B| + m_{k+1},$$

and by the induction hypothesis,

$$|B| = \sum_{i=1}^{k} |A_i| = \sum_{i=1}^{k} m_i.$$

Thus

$$|A| = \sum_{i=1}^{k} |A_i| + |A_{k+1}| = \sum_{i=1}^{k+1} |A_i| = \sum_{i=1}^{k+1} m_i.$$

Thus the result holds for $k+1$, and, by the principle of mathematical induction, it holds for all natural numbers n. □

To introduce the next principle, we consider a simple example. Suppose you are decorating cutout cookies. You have three shapes (trees, stars, and bells) and four colors of icing (red, green, yellow, blue). How many different types of cookie can you make? This problem is not hard, for you can list all twelve possibilities:

red tree	green tree	blue tree	yellow tree
red star	green star	blue star	yellow star
red bell	green bell	blue bell	yellow bell

Because there are three possibilities for the shape and four possibilities for the icing color, there are $3 \cdot 4 = 12$ possible cookie types. This example illustrates the multiplication principle.

Multiplication Principle. Suppose a set S consists of all possible distinct elements possessing two independent properties. If there are m possibilities for the first property and n possibilities for the second, then there are mn elements of S.

The multiplication principle is essentially equivalent to the statement that, for finite sets A and B,

$$|A \times B| = |A||B|. \tag{5.1}$$

We prove (5.1) using the general addition principle. Suppose A has m elements $a_1, a_2, \ldots, a_m$ and B has n elements. Observe that

$$A \times B = \bigcup_{i=1}^{m} \{(a_i, b) : b \in B\}$$

is a partition of $A \times B$. Each set in the partition has n elements and there are m sets in the partition. Thus

$$|A \times B| = \sum_{i=1}^{m} n = mn = |A||B|.$$

Although the multiplication principle is stated for the case in which each element has two properties, a simple induction argument shows that it applies more generally.

Proposition 5.2 (General Multiplication Principle). *Suppose a set S consists of all possible distinct elements possessing n independent properties, $n \geq 1$. If there are m_j possibilities for the j-th property, then there are $m_1 m_2 \cdots m_n = \prod_{j=1}^{n} m_j$ elements of S.*

Exercise 5.1. A **binary sequence of length** n is a function from $\{1, 2, \ldots, n\}$ to $\{0, 1\}$, or, equivalently, an ordered list $\{a_1, a_2, \ldots, a_n\}$ in which $a_j = 0$ or $a_j = 1$ for all $1 \leq j \leq n$. Count the binary sequences of length n.

Our next example illustrates how the addition and multiplication principles can be used in conjunction to solve more sophisticated counting problems.

Example 5.1. How many 5-letter strings start with either STR- or TH-?

We make explicit our use of the addition and multiplication principles. Let A be the set of 5-letter strings starting STR- and let B be the set of 5-letter strings starting TH-. Clearly A and B are disjoint. By the multiplication principle, A has 26^2 elements (because there are 26 possibilities for each of the fourth and fifth letters) and B has 26^3 elements. Thus by the addition principle, their union has $26^2 + 26^3$ elements.

Exercise 5.2. A coin is flipped 10 times, producing a sequence of 10 Hs and Ts. Count the number of sequences beginning with 5 Hs or 5 Ts.

5.1.2 Permutations and combinations

In this section, we develop some more sophisticated counting strategies by considering some examples. The fundamental tool is still the multiplication principle.

Example 5.2. Five friends go to a concert. They have reserved five seats in a row. How many seating arrangements are possible?

We simply apply the multiplication principle; there are five possibilities for the first open seat. Once a person has been assigned to that seat, there are only four possibilities for the second open seat. With these two assigned, there are three possibilities for the third open seat, then two for the fourth, and then the one remaining person sits in the one remaining seat. By the multiplication principle, there are $5 \cdot 4 \cdot 3 \cdot 2 \cdot 1 = 5! = 120$ possible seating arrangements.

In the preceding example, we have five distinct objects (people), each of which must be assigned to one of five different positions (seats in a concert hall). Expressed another way, we are finding the number of distinct arrangements, or *permutations*, of the five people.

Definition 5.3. Given n distinct objects, a **permutation** is an arrangement of the n objects.

Exercise 5.3. Prove that there are $n!$ permutations of n objects.

More generally, if we have n distinct objects and $m \leq n$, an m**-permutation** is an arrangement of m of the n objects. The number of m-permutations of n objects is

$$P(n, m) = n \cdot (n-1) \cdot (n-2) \cdot \ldots \cdot (n-(m-1)) = \frac{n!}{(n-m)!}. \tag{5.2}$$

Exercise 5.4. Prove (5.2).

Exercise 5.5. Suppose you have 10 distinct Valentines and you wish to send one to each of your four best friends. In how many ways can this task be done?

Sometimes given n objects and $m \leq n$, we are simply interested in the number of ways of choosing a subset consisting of m of the n objects rather than in the number of ways of arranging the objects.

Example 5.3. A math department has 12 faculty members, five of whom must be chosen to serve on the curriculum committee. It is an egalitarian sort of department, so the five committee members will be chosen at random, and, once on the committee, all members will have equal roles. How many distinct committees are possible?

If we were to use the above method, we'd get $\frac{12!}{7!}$. This method over-counts, for it treats the committee Ahern, Baouendi, Catlin, D'Angelo, Ebenfelt as different from the committee Baouendi, Ahern, D'Angelo, Catlin, Ebenfelt, and so on. Thus we must divide $\frac{12!}{7!}$ by the number of arrangements of the five committee members, which is 5!. Thus we have only $\frac{12!}{7!5!} = 792$ distinct committees.

The above example illustrates a *combination* of n objects taken m at a time. Given n distinct objects, the number of ways of selecting a set of size $m \leq n$ is

$$\binom{n}{m} = \frac{n!}{m!(n-m)!}. \tag{5.3}$$

It is the number of ways of taking m of the n objects to go in each of m positions divided by the number of ways to arrange the m objects. The expression $\binom{n}{m}$ is read "n choose m." We will see these numbers often.

Exercise 5.6. If your spice rack has 25 (complementary) spices, in how many ways can you choose six of them for your curry?

We consider one final kind of counting problem.

Example 5.4. We have four sheep in a large rectangular enclosure. We wish to add two pieces of fencing extending between opposite sides to make three parallel enclosures. In how many different ways can the sheep be enclosed if one is allowed to have one or more empty sheep enclosures?

A useful way to think about this problem is to think of it as a problem of having *six* objects–four sheep and two pieces of fence. Denote sheep by dots and fences by vertical bars. One could have any of the following:

$$\bullet\bullet|\bullet|\bullet$$
$$\bullet|\bullet\bullet|\bullet$$
$$||\bullet\bullet\bullet\bullet.$$

In the first two examples, each enclosure has at least one sheep, whereas in the final example, all four sheep are in the last enclosure. We have six possible positions and must assign fences to two of them. Such an assignment can be made in $\binom{6}{2} = 15$ ways.

Exercise 5.7. Suppose we wish to count the number of triples of non-negative integers that are solutions to $x+y+z=4$. We claim this problem is essentially the same as the problem of sheep and fences. Explain why.

5.1.3 Combinatorial identities

A *combinatorial identity* is an identity involving combinatorial objects such as permutations and combinations. Usually a combinatorial identity can be proved in several ways. One way involves the definition of permutations or combinations as ratios of factorials and copious tedious algebra. Often a more satisfying proof is possible using principles of counting. We tend to prefer the latter type of proof because it shows not only that the identity is true but also gives insight into the expressions involved. We illustrate here with some of the most famous combinatorial identities. In each case, we invite you to attempt an algebraic proof.

Counting two ways. This technique gives a surprisingly powerful method for establishing relationships that may otherwise be hard to penetrate. We consider two famous examples.

Example 5.5 (The Chairperson Identity)**.** We claim

$$n\binom{n-1}{k-1} = k\binom{n}{k}. \tag{5.4}$$

Indeed, consider the number of ways of choosing a committee of k from a set of n people with one committee member designated as chairperson. We count the number of possible committees two ways.

First, of the n people, we first choose one person to be chairperson. This choice can be made in n ways. Then, from the remaining $n-1$ people, we choose $k-1$ additional people to fill the committee. The number of possible committees is thus

$$n\binom{n-1}{k-1}.$$

As a second method, we could first choose k of n people for the committee, and then from those k, choose one to be chair. The number of committees obtained this way is

$$\binom{n}{k}k.$$

Equating these two expressions gives the identity.

Example 5.6 (The Summation Identity)**.** We claim that for $n, k \in \mathbb{N}$,

$$\sum_{i=k}^{n}\binom{i}{k} = \binom{n+1}{k+1}. \tag{5.5}$$

Sometimes the sum is taken from 0 to n instead of from k to n. The identity remains the same; we define $\binom{i}{k}$ to be zero if $i < k$ because in that case there are no ways to select a set of size k from a set of size i.

We prove (5.5) by counting in two ways the number of binary sequences of length $n+1$ with $k+1$ ones. For the first method, we simply note that there are $\binom{n+1}{k+1}$ such sequences, for we must choose $k+1$ of the $n+1$ positions into which to place a one and then the remaining elements of the sequence are zeros.

Second, we count the sequences according to the position of the right-most one. This one could be in any position from $k+1$ to $n+1$. If it is in position $k+1$, the remaining k ones are in positions 1 through k, and there is only $1 = \binom{k}{k}$ way to choose their positions. If the right-most one is in position $k+1+1$, then the remaining k ones are all in the $k+1$ positions to the left of the right-most one, and there are $\binom{k+1}{k}$ ways to choose their positions. In general, if the right-most one is in position $i+1$ for $k \leq i \leq n$, there are $\binom{i}{k}$ ways to choose the positions of the remaining k ones from the i available positions. Counting this way, the number of possible binary sequences of length $n+1$ with $k+1$ ones is $\sum_{i=k}^{n}\binom{i}{k}$. Equating our two expressions gives the Summation Identity.

Exercise 5.8. Show that $n^2 = 2\binom{n}{2} + n$ by counting in two ways the ordered pairs (j, k) of natural numbers with $1 \leq j, k \leq n$. Also give an algebraic proof.

The Binomial Theorem. You may know the following theorem from high school algebra.

Theorem 5.1 (Binomial Theorem). *Let x and y be real numbers and let n be a non-negative integer. Then*

$$(x+y)^n = \sum_{k=0}^{n} \binom{n}{k} x^{n-k} y^k. \tag{5.6}$$

To make sure we understand the statement of the theorem, let us see what it says for $n = 1, 2, 3$. When $n = 1$,

$$\begin{aligned} \sum_{k=0}^{1} \binom{1}{k} x^{1-k} y^k &= \binom{1}{0} x^{1-0} y^0 + \binom{1}{1} x^{1-1} y^1 \\ &= x + y \\ &= (x+y)^1. \end{aligned}$$

When $n = 2$,

$$\begin{aligned} \sum_{k=0}^{2} \binom{2}{k} x^{2-k} y^k &= \binom{2}{0} x^{2-0} y^0 + \binom{2}{1} x^{2-1} y^1 + \binom{2}{2} x^{2-2} y^2 \\ &= x^2 + 2xy + y^2 \\ &= (x+y)^2. \end{aligned}$$

Finally, when $n = 3$,

$$\begin{aligned} \sum_{k=0}^{3} \binom{3}{k} x^{3-k} y^k &= \binom{3}{0} x^{3-0} y^0 + \binom{3}{1} x^{3-1} y^1 + \binom{3}{2} x^{3-2} y^2 + \binom{3}{3} x^{3-3} y^3 \\ &= x^3 + 3x^2 y + 3xy^2 + y^3 \\ &= (x+y)^3. \end{aligned}$$

These first few examples give us the confidence we need to embark on the proof.

Proof. An induction proof is possible; see Exercise 5.10. We give a different proof.

Observe that $(x+y)^n$ is a polynomial in x and y of degree n. Consider the coefficient of $x^{n-k}y^k$ for $0 \leq k \leq n$. This monomial arises in the product by choosing y as k of the n factors and x as the other $n-k$. There are $\binom{n}{k}$ ways to choose the k factors of y, and thus $x^{n-k}y^k$ appears $\binom{n}{k}$ times in the product, each time with coefficient 1. Combining like terms gives the result. □

Exercise 5.9. Consider the identity

$$\binom{n}{k} = \binom{n-1}{k-1} + \binom{n-1}{k} \tag{5.7}$$

for $n \geq 1$ and $1 \leq k \leq n$.

(a) Prove (5.7) using algebra and the explicit formula $\binom{a}{b} = \frac{a!}{b!(a-b)!}$.

(b) Prove (5.7) using the Binomial Theorem and the polynomial identity

$$(x+y)^n = (x+y)^{n-1}(x+y).$$

Exercise 5.10. Prove the Binomial Theorem using induction and part (a) of Exercise 5.9.

Because of their appearance in the Binomial Theorem, the numbers $\binom{n}{k}$ are called *binomial coefficients.* As we have already seen, binomial coefficients satisfy many beautiful identities. The problems ask you to explore several more. We close this section by looking at one of the most famous properties of binomial coefficients. We will use them to form *Pascal's Triangle.* Pascal's Triangle has rows numbered with the non-negative integers $0, 1, 2, \ldots$. Row n contains the $n+1$ binomial coefficients $\binom{n}{k}$ for $0 \leq k \leq n$. By the Binomial Theorem, we can also think of row n as containing the coefficients in the standard form of the expansion of $(x+y)^n$. Here are the first four rows of Pascal's Triangle.

$$\begin{array}{ccccccc} & & & 1 & & & \\ & & 1 & & 1 & & \\ & 1 & & 2 & & 1 & \\ 1 & & 3 & & 3 & & 1 \end{array}$$

To get the next row, we could, of course, simply evaluate all the binomial coefficients $\binom{4}{k}$ for $0 \leq k \leq 5$. Alternatively, we could use (5.7), which is a recursive formula expressing the binomial coefficients in row n in terms of binomial coefficients in row $n-1$. For example,

$$\binom{4}{1} = \binom{3}{0} + \binom{3}{1} = 1 + 3 = 4,$$

and

$$\binom{4}{2} = \binom{3}{1} + \binom{3}{2} = 3 + 3 = 6.$$

We see that the row for $n = 4$ has entries $1, 4, 6, 4, 1$.

Exercise 5.11. Use the recursive formula (5.7) to find the rows in Pascal's Triangle for $n = 5, 6$, and 7.

5.2 Linear Recurrence Relations

Definition 5.4. A **recurrence relation** for a sequence $\{a_n\}$ is an expression that relates n and the values $a_0, \ldots, a_n$. If the recurrence relates a_n just to n and the k previous values $a_{n-1}, a_{n-2}, \ldots, a_{n-k}$, we call it a k-th order recurrence relation.

We have already seen some simple recurrence relations. For example, in Chapter 2 we considered sequences defined recursively by

$$\begin{aligned} T_0 &= 0, \qquad & T_n &= T_{n-1} + n \\ c_0 &= c, \qquad & c_n &= c_{n-1} + d \\ a_0 &= a, \qquad & a_n &= ra_{n-1}. \end{aligned}$$

These are all first-order recurrence relations because the n-th term depends only on the one preceding term. In each of these cases, we are able to iterate to "solve" the recurrence relation, i.e., to obtain an explicit n-th term formula. For example, the solution to the third recurrence relation above is the sequence with n-th term given by

$$a_n = ar^n, \quad n \geq 0. \tag{5.8}$$

In this section, we develop a method for solving some more complicated recurrence relations. While our method will only apply to a small class of recurrence relations (the linear, homogeneous, constant-coefficient recurrence relations), we will see that this class includes many interesting examples.

5.2.1 An example

Before stating and proving a general result, let us look at a more complicated example.

Example 5.7. Let $a_0 = 2$, $a_1 = 7$, and suppose that for $n \geq 2$,

$$a_n = 7a_{n-1} - 12a_{n-2}. \tag{5.9}$$

Although this recurrence relation is more complicated than $a_n = ra_{n-1}$, we might still hope for a solution of a similar form, i.e., we might hope for non-zero a and r for which the sequence with n-th term $a_n = ar^n$ is a solution to (5.9). Then for $n \geq 2$ we would have $a_{n-1} = ar^{n-1}$, $a_{n-2} = ar^{n-2}$, and (5.9) would read

$$ar^n = 7ar^{n-1} - 12ar^{n-2}.$$

Dividing out ar^{n-2} and rearranging gives the equation

$$r^2 - 7r + 12 = 0,$$

which has solutions $r = 3, 4$.

Now, if the sequences with n-th term $a_n = A(3)^n$ and $a_n = B(4)^n$ are solutions, so is their sum (see Exercise 5.12). Thus the sequence given by

$$a_n = A(3)^n + B(4)^n$$

is a solution. Because we are given $a_0 = 2$ and $a_1 = 7$, we can solve for A and B. We obtain the system

$$\begin{aligned} A + B &= 2 \\ 3A + 4B &= 7 \end{aligned}$$

with solution $A = B = 1$.

Exercise 5.12. Consider the recurrence relation

$$a_n = ca_{n-1} + da_{n-2}, \quad n \geq 2$$

for constants c and d. Suppose $\{a_n\}$ and $\{b_n\}$ are two sequences satisfying the recurrence relation. Show that $\{Aa_n + Bb_n\}$ also satisfies the recurrence.

Exercise 5.13 (Fibonacci sequence)**.** Use the method of the example to solve

$$F_0 = 0, \quad F_1 = 1, \quad \text{and} \quad F_n = F_{n-1} + F_{n-2} \quad \text{for} \quad n \geq 2.$$

The method we have used for solving second-order linear, homogeneous recurrence relations with constant coefficients might look familiar. We use the same method when we solve second-order, linear, homogeneous differential equations with constant coefficients. For example, consider the differential equation

$$\frac{d^2y}{dx^2} - 7\frac{dy}{dx} + 12y = 0$$

for y a twice-differentiable function of x. In order to solve this equation, we suppose that there is a solution of the form $y(x) = e^{rx}$ (for the simpler differential equation $\frac{dy}{dx} = ry$ would have a solution of this form). Substituting into the differential equation gives

$$r^2e^{rx} - 7re^{rx} + 12e^{rx} = 0.$$

Dividing by e^{rx} again yields the quadratic equation $r^2 - 7r + 12 = 0$ with solutions $r = 3, 4$. We conclude that the general solution to the differential equation is $Ae^{3x} + Be^{4x}$. If we have initial conditions, we can solve for A and B.

If you have never studied differential equations, don't worry; we include these remarks to help those who have taken such a course make connections among the courses they have taken. If these ideas are new to you, the message to take away is that often the same method of solution can be used for problems from different areas of mathematics.

5.2.2 General results

In the previous subsection, we illustrated a method for solving a second-order, linear, homogeneous recurrence relation with constant coefficients, but we did not prove that such an approach always works. In this subsection, we establish several results that address questions of existence and uniqueness of solutions.

Proposition 5.3. *A set of j initial values $a_0, \ldots, a_{j-1}$ and a j-th order recurrence relation for a_n for $n \geq j$ uniquely determine the terms of a sequence $\{a_n\}$.*

Proof. The proof is by strong induction on n. For the basis step, we need only note that the first j terms of the sequence are given explicitly and so are obviously uniquely determined.

Now suppose $k \geq j-1$ and that all a_i for $0 \leq i \leq k$ are uniquely determined. Consider $k+1$. Because $k+1 \geq j$, the value of a_{k+1} is given by the j-th order recurrence relation, i.e., it depends uniquely on the values $a_k, a_{k-1}, \ldots, a_{k-j+1}$. Because, by the inductive hypothesis, the numbers $a_k, a_{k-1}, \ldots, a_{k-j+1}$ are uniquely determined, a_{k+1} is uniquely determined. Thus the result holds for $k+1$, and by the principle of mathematical induction, it holds for all n. □

Theorem 5.2. *Let a, b, c, and d be constants. Suppose $a_0 = a$, $a_1 = b$, and*

$$a_n = ca_{n-1} + da_{n-2}, \quad n \geq 2. \tag{5.10}$$

Consider the quadratic equation $x^2 = cx + d$.

(a) If this equation has unique solutions α and β, then there exist constants A and B such that $a_n = A\alpha^n + B\beta^n$ for all n.

(b) If this equation has a single (repeated) solution α, then there exist constants A and B such that $a_n = A\alpha^n + Bn\alpha^n$ for all n.

Proof. We first consider part (a). Suppose $x^2 = cx + d$ has distinct solutions α and β. Then

$$\alpha^2 = c\alpha + d.$$

If we multiply this equation by α^{n-2} for $n \geq 2$, we obtain

$$\alpha^n = c\alpha^{n-1} + d\alpha^{n-2}.$$

Thus for $n \geq 2$, the sequence $\{\alpha^n\}$ satisfies (5.10). An identical argument shows that, for $n \geq 2$, the sequence $\{\beta^n\}$ satisfies (5.10). Then by Exercise 5.12, for any A, B, the sequence $\{A\alpha^n + B\beta^n\}$ satisfies (5.10) for $n \geq 2$. In order to complete the proof of part (a), we need only show that the values of A and B can now be chosen so that

$$\begin{aligned} A\alpha^0 + B\beta^0 &= a_0 = a \\ A\alpha^1 + B\beta^1 &= a_1 = b. \end{aligned}$$

We find that $B = \frac{b-a\alpha}{\beta-\alpha}$ and $A = \frac{a\beta-b}{\beta-\alpha}$.

Next we consider part (b). In this case, the single solution α to $x^2 = cx + d$ satisfies

$$(x-\alpha)^2 = 0,$$

or, equivalently,

$$x^2 = 2\alpha x - \alpha^2.$$

Thus $c = 2\alpha$ and $d = -\alpha^2$, which implies $c\alpha + 2d = 0$. Now, as above,

$$\alpha^2 = c\alpha + d,$$

and multiplying through by α^{n-2} for $n \geq 2$ shows that the sequence $\{\alpha^n\}$ satisfies the recurrence for $n \geq 2$. Alternatively, we may multiply through by $n\alpha^{n-2}$ to obtain

$$\begin{aligned} n\alpha^n &= cn\alpha^{n-1} + dn\alpha^{n-2} \\ &= c(n-1)\alpha^{n-1} + d(n-2)\alpha^{n-2} + c\alpha^{n-1} + 2d\alpha^{n-2} \\ &= c(n-1)\alpha^{n-1} + d(n-2)\alpha^{n-2} + \alpha^{n-2}(c\alpha + 2d) \\ &= c(n-1)\alpha^{n-1} + d(n-2)\alpha^{n-2}. \end{aligned}$$

Thus the sequence $\{n\alpha^n\}$ satisfies the recurrence for $n \geq 2$. By Exercise 5.12, any sequence of the form $\{A\alpha^n + Bn\alpha^n\}$ satisfies the recurrence for $n \geq 2$. We will have completed the proof if we find A and B so that the initial conditions are satisfied. We leave this last step to the reader as the next exercise. □

Exercise 5.14. Find A and B so that $\{A\alpha^n + Bn\alpha^n\}$ satisfies the initial conditions $a_0 = a$ and $a_1 = b$.

Exercise 5.15. Solve the recurrence relation $a_n = 6a_{n-1} - 9a_{n-2}$, $n \geq 2$ with initial conditions $a_0 = 2$ and $a_1 = 9$.

We can use an analogous method to solve higher-order, linear, homogeneous, constant coefficient recurrence relations. As we did in the second-order case, we assume a solution of the form $a_n = ar^n$. Substituting into the recurrence relation again leads to a polynomials equation, and our goal is to find its roots r_i. If a root has multiplicity $j > 1$, in addition to the solution r_1^n, we also obtain solutions of $n^l r_i^n$ for $1 \leq l \leq j-1$. We omit the formal statement of this theorem and its proof because there are no new ideas here, but considerable cumbersome notation.

5.3 Analysis of Algorithms

5.3.1 Some simple algorithms

Roughly speaking, an algorithm is a finite sequence of precise steps that, when applied to some general input, yields a unique output. You are already familiar with many algorithms. For example, in grade school we all learn a long division algorithm; given a non-negative dividend n and a divisor d, the algorithm tells us how to write $n = qd + r$ (with $0 \le r < d$) in a finite number of steps. Each step can be described precisely, and each step produces a unique output. Of course, the long division algorithm we learn in school is not the only way to accomplish division of natural numbers, and if one's only goal is to know the answer to a certain division problem, it doesn't much matter how one gets the answer. The advantages of an algorithm are its generality and its precision; it applies to a wide range of inputs, and we can easily write a program in our favorite programming language to execute the algorithm.

We study algorithms in an Introduction to Proof book for several reasons. First, whenever we devise an algorithm to do something, we need to prove that it does what it is supposed to do. Second, the kind of thinking that we use when we devise an algorithm is the same kind of orderly and precise thinking we aim for when we write mathematical arguments. Third, the analysis of the complexity of algorithms introduces some notions that are used throughout mathematics.

In this subsection, let us look at some very simple algorithms. In each case, we will want to pay attention to the number of steps in the algorithm as a function of the input.

An algorithm to find the largest element in a list. Suppose we have n numbers $a_0, a_1, \ldots, a_{n-1}$ and we want to find the largest. Here is a simple algorithm that accomplishes this task.

```
# Finds largest element of a finite list

def largest_element(List):
    Largest=List[0] #initializes Largest to be the first element
        of the list

    for i in range(1,len(List)):
        if List[i]>Largest:
            Largest=List[i]

    return(Largest)
```

Notice that we go through the loop $n-1$ times, each time comparing *List*[i] to the stored value of *Largest*. Each time, if *List*[i] $>$ *Largest*, we update *Largest*.

Exercise 5.16. Consider the list $S = [1, 8, 9, 4, 6]$. Clearly its largest element is 9. Show how the largest element algorithm works on this sequence by describing what happens each time through the loop.

A simple sorting algorithm. Suppose we want to do more. Suppose we want to take a list and rearrange it so that the elements are ordered from largest to smallest. We consider the most elementary approach. We start with an n-element list S we want to order and an empty list T. We use the largest element function above to find the largest element of S. We append this element to T and remove it from S, leaving a list with $n-1$ elements. We now apply the largest element function to find the largest element of the new list. This element is, of course, the second largest element of the original list S. We append this element to T and remove it from our list to be ordered. Our updated list now has $n-2$ elements in it, and our updated T has 2 elements. We repeat this process until our list to be ordered is empty. Here is Python code for this algorithm.

```
# Input: a list S. Output: list reordered from largest to
    smallest

# Finds the largest element of any List
def largest_element(List):
    Largest=List[0] #initializes Largest to be the first element
        of the list

    for i in range(1,len(List)):
        if List[i]>Largest:
            Largest=List[i]
    return(Largest)

# Uses the largest element function to reorder the list
def ordered_list(S):
    List=S #initialize List to be the input list S
    T=[] #defines an empty list to hold reordered list
    while len(List)>0:
        T.append(largest_element(List)) #updates T by appending
            the largest element of the current List
        List.remove(largest_element(List)) #updates List,
            removing the largest element
    return(T)
```

We now consider the complexity of this sorting algorithm. We consider the number of comparisons that must be made by the largest element function. Suppose we start with the n-element list S_0. Finding its largest element requires us to make $n-1$ comparisons. Next we apply the largest element function to the $(n-1)$-element list S_1 obtained by taking out of S_0 its largest element. This step requires another $n-2$ comparisons. In general, at step k we apply the largest element function to the $(n-k+1)$-element list S_{k-1}. This step requires $n-k$ comparisons. Thus the total number of comparisons required for our sorting algorithm is

$$(n-1)+(n-2)+\ldots+(n-k)+\ldots+2+1=\frac{n(n-1)}{2}=\frac{1}{2}n^2-\frac{1}{2}n.$$

Exercise 5.17. Consider the list $S=[1,8,9,4,6]$. Show how the sorting algorithm works on this list by describing the contents of *List* and T before and after each pass through the *while* loop.

5.3.2 O, Ω, and Θ notation

Of course we analyze the number of steps in an algorithm to know how much computing time it will take. For such an analysis, it matters if one algorithm takes n^2 steps and another takes $n!$ steps (the latter being *much* larger for large n), but it does not really matter if one algorithm takes n^2 steps and another takes $n^2 + 3$ steps or even $2n^2$ steps. In this subsection we discuss ways to capture the essential size of a function and ways to compare functions.

Definition 5.5. Let $f, g\colon \mathbb{N} \to \mathbb{R}$. Then $f(n) = O(g(n))$ if there exist a positive constant C and an $N \in \mathbb{N}$ such that, for all $n \geq N$, $|f(n)| \leq C|g(n)|$.

Definition 5.6. Let $f, g\colon \mathbb{N} \to \mathbb{R}$. Then $f(n) = \Omega(g(n))$ if there exist a positive constant C and an $N \in \mathbb{N}$ such that, for all $n \geq N$, $|f(n)| \geq C|g(n)|$.

Definition 5.7. Let $f, g\colon \mathbb{N} \to \mathbb{R}$. Then $f(n) = \Theta(g(n))$ if $f(n) = O(g(n))$ and $f(n) = \Omega(g(n))$.

This last definition says that, except for finitely many n, f is bounded above and bounded below by g. Indeed, because $f(n) = O(g(n))$, there exist $C > 0$ and $N_1 \in \mathbb{N}$ such that

$$|f(n)| \leq C|g(n)| \quad \text{for all } n \geq N_1.$$

Furthermore, because $f(n) = \Omega(g(n))$, there exist $D > 0$ and $N_2 \in \mathbb{N}$ such that

$$|f(n)| \geq D|g(n)| \quad \text{for all } n \geq N_2.$$

Thus if $n \geq N = \max\{N_1, N_2\}$, both inequalities hold:

$$D|g(n)| \leq |f(n)| \leq C|g(n)| \quad \text{for all } n \geq N.$$

Let's look at some examples.

Example 5.8. We saw that the algorithm that finds the largest element in a sequence of length n takes $s(n) = n - 1$ steps. We will prove $s(n) = \Theta(n)$.

First we prove the easy direction. For all $n \in \mathbb{N}$,

$$\begin{aligned} |s(n)| &= n - 1 \\ &\leq n \\ &= |n|. \end{aligned}$$

Thus $s(n) = O(n)$.

For the other direction, observe that for $n \geq 2$, $1 \leq \frac{n}{2}$. Thus for $n \geq 2$,

$$\begin{aligned} |s(n)| &= n - 1 \\ &\geq n - \frac{n}{2} \\ &= \frac{1}{2}|n|, \end{aligned}$$

and so $s(n) = \Omega(n)$. Thus $s(n) = \Theta(n)$.

Example 5.9. In our sorting algorithm above, we determined that the number of times through the comparison loop is

$$s(n) = \frac{1}{2}n^2 - \frac{1}{2}n.$$

We will show that $s(n) = \Theta(n^2)$.

Observe first that for all $n \geq 1$,

$$\begin{aligned} |s(n)| &= \frac{1}{2}n^2 - \frac{1}{2}n \\ &\leq \frac{1}{2}n^2. \end{aligned}$$

Thus $s(n) = O(n^2)$. Also, because $n \leq \frac{1}{2}n^2$ for all $n \geq 2$, for such n

$$\begin{aligned} |s(n)| &= \frac{1}{2}n^2 - \frac{1}{2}n \\ &\geq \frac{1}{2}n^2 - \frac{1}{2} \cdot \frac{1}{2}n^2 \\ &= \frac{1}{4}n^2. \end{aligned}$$

Therefore $s(n) = \Omega(n^2)$. We conclude that $s(n) = \Theta(n^2)$.

Exercise 5.18. Let a be any positive constant.

(a) Show that $an = O(n^2)$.

(b) Negate the definition of $f(n) = O(g(n))$ and show that $n^2 = O(n)$ is false.

Exercise 5.19. Let b be any constant and let a be any non-zero constant. Show that $n + b = \Theta(n)$ and that $an + b = \Theta(n)$.

5.3.3 Analysis of the binary search algorithm

We end this section by analyzing one final simple algorithm. Here is the problem: We have a (possibly very long) list of numbers arranged in order from smallest to largest and we want to determine whether a specific number, called the *key*, appears somewhere on the list. The most obvious thing to do would be to compare each element of the list to the key and to see if there is a match. If our list has n elements, in the worst-case scenario, this algorithm requires n comparisons. It is possible, however, to come up with more efficient algorithms. We describe a simple one known as a *binary search*.

The binary search algorithm is an example of a divide-and-conquer algorithm. We will see several such algorithms in the second half of this book. In this algorithm, we begin by dividing our initial list (nearly) in half. We test to see whether the key equals the element at this dividing point. If it does, the process terminates. If not, because the list is ordered, we can compare the key with this element to determine whether the element we seek would fall in the first half or second half of the list. We then recursively apply the algorithm to this new shorter list. The algorithm returns the index at which the key is located if it is on the list and it returns -1 if the key is not found on the list.

Let us look at the Python code for the algorithm. We have written it in such a way that if you run it, you can try it out on some lists of your own.

```
# input: list in increasing order and a key. output: index at
    which key occurs or -1 if key is not on the list

S=input("enter a list in increasing order: ")
key=input("enter a key: ")
i=0
j=len(S)-1

def binary_search(list,i,j,key):
    if i>j: #key not found
        return(-1)
    else:
        k=int((i+j)/2) #index of element near middle
        if list[k]==key: #key found
            return(k)
        if list[k]>key: #search first half
            return(binary_search(list,i,k-1,key))
        if list[k]<key: #search second half
            return(binary_search(list,k+1,j,key))

print(binary_search(S,i,j,key))
```

To make sure we understand the algorithm, we see what happens when the list is $S = [-2, 0, 4, 5, 7, 10, 14, 15]$ and the key is 14. Note that $i = 0$ and $j = 7$. We want to find *binary_search*$(S, 0, 7, 14)$. Because $0 > 7$ is false, we execute the *else* statement. We start by setting k equal to the greatest integer in $\frac{0+7}{2}$, which is 3. We then compare $S[3] = 5$ to the key of 14. Precisely one of the three *if* statements is true. In this case, it is the statement $S[3] < 14$. Thus we call the binary search function again, this time finding *binary_search*$(S, 4, 7, 14)$. Thus now we are only considering the portion of the list associated with the indices from 4 to 7 (i.e., the second half of the list). We update the value of k to be the greatest integer in $\frac{4+7}{2}$, which is 5. We now compare $S[5] = 10$ to the key and discover again that $S[5] < 14$. We must thus now consider *binary_search*$(S, 6, 7, 14)$. We update k to be the greatest integer in $\frac{6+7}{2}$, which is 6. Now when we do the comparison of $S[6] = 14$ with the key, we find that they are equal. We do not call the function again but rather return k. We have found that $S[6]$ is equal to the key.

Exercise 5.20. Describe how the algorithm works when $S = [2, 3, 4, 5, 6, 7, 8, 9]$ and the key is 0. Of course the function should return -1.

We claimed above that this binary search algorithm is more efficient than the simple search. To establish this claim, we will analyze the worst-case scenario for the algorithm. Clearly the worst case occurs when the key is not on the list. We will let t_n denote the number of times *binary_search* is called when the input list has length n. We will obtain a recurrence relation and initial conditions for t_n.

When $n = 1$, $i = j = 0$ and we consider *binary_search*$(S, 0, 0, key)$. Because $0 > 0$ is false, we compare $S[0]$ with the key. In the worst-case scenario, $S[0] \neq key$. Depending on the relative sizes of the two, we either call *binary_search*$(S, 0, -1, key)$ or *binary_search*$(S, 1, 0, key)$. Because both $0 > -1$ and $1 > 0$ are true, in either case the process now terminates and the function returns -1. Because we have invoked *binary_search* twice, $t_1 = 2$.

Now suppose $n > 1$ and consider *binary_search*$(S, 0, n-1, key)$. Because $n - 1 > 0$, we proceed to the *else* statement and set k equal to the greatest integer less than or equal to $\frac{n-1}{2}$. We denote this number by $\lfloor \frac{n-1}{2} \rfloor$. In the worst-case scenario, $S[k] \neq key$. Depending on the relative sizes, we now either invoke *binary_search*$(S, 0, k-1, key)$ or *binary_search*$(S, k+1, n-1, key)$. In the first case, we would be considering a list of length $k = \lfloor \frac{n-1}{2} \rfloor$, and in the second case we would be considering a list of length

$$(n-1) - (k+1) + 1 = n - 1 - k = n - 1 - \left\lfloor \frac{n-1}{2} \right\rfloor = \left\lfloor \frac{n}{2} \right\rfloor.$$

In the worst-case scenario, we must consider the list of length $\lfloor \frac{n}{2} \rfloor$, which requires $t_{\lfloor \frac{n}{2} \rfloor}$ steps. Thus

$$t_n = 1 + t_{\lfloor \frac{n}{2} \rfloor}.$$

Ideally we would like an explicit formula for t_n. Unfortunately, such a formula is not easy to find for general n. We can, however, easily obtain a formula when n is a power of 2. Because every natural number is between two powers of 2, we can get estimates on t_n for all n.

We thus suppose first that $n = 2^m$ for $m > 0$. Then

$$t_{2^m} = 1 + t_{2^{m-1}}.$$

Iterating this recurrence relation gives

$$t_{2^m} = m + t_{2^{m-m}} = m + t_1 = m + 2.$$

Next suppose n is arbitrary. Find the natural number m with $2^m \leq n < 2^{m+1}$. Observe that

$$m \leq \log_2 n < m + 1. \tag{5.11}$$

We now estimate t_n.

$$m + 2 = t_{2^m} \leq t_n < t_{2^{m+1}} = m + 3.$$

By (5.11), $m + 2 > (\log_2 n) + 1$ and $m + 3 \leq (\log_2 n) + 3$. Thus

$$(\log_2 n) + 1 < t_n < (\log_2 n) + 3.$$

It follows that $t_n = \Theta(\log_2 n)$. For large n, $\log_2 n$ is much smaller than n. We thus see that for long lists, binary search is a more efficient algorithm than the very simple algorithm that compares each element of the list to the key.

Exercise 5.21. Show that for any constant a, $(\log_2 n) + a = \Theta(\log_2 n)$.

5.4 Problems

1. Let A be a set with n elements. How many subsets does A have? (The empty set counts as a subset of A.)

2. Find and prove a simple formula for $\sum_{k=0}^{n} \binom{n}{k}$. Give as many distinct proofs as you can.

3. Find and prove a simple formula for $\sum_{k=0}^{n} (-1)^k \binom{n}{k}$.

4. Consider an $n \times n$ grid of squares. How many routes are there from the lower left corner to the upper right corner if one may only travel rightward and upward along the grid lines?

5. In how many ways can n pairs be selected from a set of $2n$ distinct elements?

6. A **monomial** in n variables is an expression of the form
$$x_1^{a_1} x_2^{a_2} \cdots x_n^{a_n}$$
for non-negative integers a_j. The **degree** of the monomial is $\sum_{j=1}^{n} a_j$. Count the number of monomials of degree d in n variables.

7. Suppose you have n dollars. Every afternoon you go to the market for a snack. Each day you either buy a soda for \$1, a pastry for \$2, or hot cocoa for \$2. Let a_n denote the number of ways to spend all n dollars.
   (a) Find a_1, a_2, a_3.
   (b) Find a recurrence relation (with initial conditions) satisfied by a_n.
   (c) Find an explicit formula for a_n.

8. Refer to Exercise 5.19. State and prove a corresponding result for quadratic polynomials. Then state and prove a result for polynomials of arbitrary degree.

9. Fix a natural number k and consider
$$s_n = \sum_{i=1}^{n} i^k = 1^k + 2^k + \ldots + n^k.$$
For what natural number(s) K is $s_n = \Theta(n^K)$ true? Prove your claim.

10. Consider the following simple algorithm.

```
i=n
while i>0:
    for j in range(0,n):
        x=x+1
    i=int(i/2)
```

   (a) For $n = 2, 4, 8, 16$, determine the number of times the line $x = x + 1$ is executed. (You can check your answer by running the code with x initially set to 0.)

   (b) Use theta notation to describe s_n, the number of times the line $x = x + 1$ is executed as a function of n. (Hint: Start by considering n a power of 2.)

Programming Project

1. Write a merge function that takes as input two lists, each in increasing order, and returns a single list in increasing order.

2. Analyze your algorithm, i.e., find an expression for t_n, the number of steps your algorithm takes when the sum of the lengths of the input lists is n.

Chapter 6

Number Systems and Algebraic Structures

We have already encountered many number systems, including the natural numbers, the integers, the rational numbers, and the real numbers. In our first treatment, we took all these number systems as familiar and did not indicate how they are rigorously constructed. We have enough background that we could, if we wanted, rigorously construct the natural numbers, then the integers, then the rational numbers, then the real numbers. We will omit the first step of this construction, the rigorous construction of the natural numbers, because we don't want to be overly pedantic in our dealings with a number system of which we already have a deep intuitive understanding. In this chapter we will, however, discuss properties of the natural numbers and integers in more detail, and we will discuss the construction of the rational numbers from the integers. In the second half of the book we will discuss the construction of the real numbers from the rational numbers.

Each of the familiar number systems we study has operations on it satisfying certain useful properties such as commutativity and associativity. Because many different mathematical systems have operations satisfying similar properties, it makes sense to study these lists of properties independently of the specific examples where we first met them. Such a study of *algebraic structures* is the aim of the second half of the chapter.

6.1 Representations of Natural Numbers

You go to the store and buy two standard cartons of eggs. Visualize the eggs. Now take out a piece of paper and write down the number of eggs you have. Unless you have not shopped for eggs recently or you are being intentionally non-conformist, you wrote 24 on your paper. We have all been taught to think and do our arithmetic in base 10; given a natural number n, we write it as

$$n = \sum_{j=0}^{k} d_j 10^j, \quad 0 \le d_j \le 9 \tag{6.1}$$

and then represent n as the ordered string of digits $d_k d_{k-1} \ldots d_1 d_0$. In the case of the eggs, the number n of eggs is

$$n = 2 \cdot 10^1 + 4 \cdot 10^0,$$

and thus we write $n = 24$.

Other bases are possible. Given how eggs are packaged, 12 might be a natural base. Our number of eggs is

$$n = 2 \cdot 12^1 + 0 \cdot 12^0.$$

We write $n = 20_{12}$ in this case to indicate that n has been expressed in a base other than the standard. Other representations are

$$n = 4 \cdot 6^1 + 0 \cdot 6^0 = 40_6$$
$$n = 1 \cdot 2^4 + 1 \cdot 2^3 + 0 \cdot 2^2 + 0 \cdot 2^1 + 0 \cdot 2^0 = 11000_2.$$

In fact any natural number other than 1 may be used as a base.

Definition 6.1. Let b be a natural number greater than 1. Let n be any natural number. Its **base b representation** is $(d_k d_{k-1} \ldots d_1 d_0)_b$ if

$$n = \sum_{j=0}^{k} d_j b^j, \quad 0 \le d_j \le b - 1. \tag{6.2}$$

The d_j are called the **digits** of the base b representation of n.

When the base is 2, we refer to the *binary representation* of n.

Exercise 6.1. Write the base 10, base 3, and base 2 representations of your age. List all bases b (if any) between 2 and 10 in which the base b representation of your age has only a single non-zero digit.

Exercise 6.2. Explain the following T-shirt quote: "There are 10 kinds of people in the world–those who understand binary and those who don't."

Implicit in the above definition is the theorem that, given a natural number $b > 1$, every natural number n *has* a base b representation and this representation is *unique*. We will prove this statement. How? We use an approach that illustrates one of the main themes of the book: If we can figure out an algorithm to do something, we are most of the way to a valid proof that it is possible. We're not saying that this approach is the only or best way to find proofs, only that it is a valuable tool for all creators of mathematics.

6.1.1 Developing an algorithm to convert a number from base 10 to base 2

In the previous subsection, we looked at the representations of some small natural numbers (such as your age and the number of eggs in two standard cartons) in different bases. There, we did not explicitly give an algorithm for expressing a given natural number in terms of a fixed base b. Developing such an algorithm is the goal of this section. We start with a few examples.

Example 6.1. Let us express 148 in base 3. Thus we wish to find digits $0 \leq d_j \leq 2$ such that

$$148 = d_k 3^k + \ldots + d_3 3^3 + d_2 3^2 + d_1 3 + d_0, \tag{6.3}$$

with $d_k \neq 0$. Let's first figure out what k is. If (6.3) is to be a valid base 3 representation of 148, we must have

$$3^k \leq 148,$$

with equality if and only if $d_k = 1$ and all other digits are zero. At the other extreme, each digit in (6.3) could be as large as 2. Using the formula for a finite geometric series, we find

$$148 \leq \sum_{j=0}^{k} 2 \cdot 3^k = 3^{k+1} - 1 < 3^{k+1}.$$

By checking successive powers of 3, we find

$$3^4 = 81 \leq 148 < 243 = 3^5.$$

Thus $k = 4$. We next figure out $d_k = d_4$. d_4 must satisfy

$$d_4 3^4 \leq 148 \quad \text{but} \quad (d_4 + 1)3^4 > 148.$$

Because

$$1 \cdot 3^4 = 81 \leq 148 < 162 = 2 \cdot 3^4,$$

$d_4 = 1$. We now know that

$$148 = 1 \cdot 3^4 + d_3 \cdot 3^3 + d_2 \cdot 3^2 + d_1 \cdot 3 + d_0.$$

Furthermore, because

$$148 - 1 \cdot 3^4 = 67 = d_3 \cdot 3^3 + d_2 \cdot 3^2 + d_1 \cdot 3 + d_0,$$

we have reduced the problem of finding the base 3 representation of 148 to the problem of finding the base 3 representation of the smaller number 67.

We now repeat the process with 67; we determine the first non-zero digit in its base 3 representation. In this case,

$$3^3 = 27 \leq 67 < 3^4,$$

so the first non-zero digit is, in fact, d_3. Furthermore,

$$2 \cdot 3^3 = 54 \leq 67 < 81 = 3 \cdot 3^3,$$

so $d_3 = 2$. Thus

$$67 = 2 \cdot 3^3 + d_2 3^2 + d_1 3 + d_0.$$

It thus suffices to find the base 3 representation for $67 - 54 = 13$. Because

$$3^2 \leq 13 < 27 = 3^3,$$

d_2 is non-zero. It is, in fact, equal to 1 because

$$1 \cdot 3^2 = 9 \leq 13 < 18 = 2 \cdot 3^2.$$

We now know

$$13 - 1 \cdot 3^2 = 4 = 1 \cdot 3 + 1.$$

Putting together all of this information we obtain

$$148 = 1 \cdot 3^4 + 2 \cdot 3^3 + 1 \cdot 3^2 + 1 \cdot 3 + 1,$$

i.e., $148 = 12111_3$.

Example 6.2. We do a second example, this time converting to base 2. Let us again begin with the number $n = 148$. We wish to write

$$148 = \sum_{j=0}^{k} d_j 2^j.$$

k must satisfy

$$2^k \leq 148 \leq \sum_{j=0}^{k} 2^j = 2^{k+1} - 1 < 2^{k+1}.$$

We observe that

$$2^7 = 128 \leq 148 < 256 = 2^8.$$

Thus $k = 7$, and necessarily $d_k = 1$. We now consider

$$148 - 2^7 = 148 - 128 = 20.$$

We easily see

$$2^4 = 16 \leq 20 < 32 = 2^5,$$

and so 20 has d_4 as its first non-zero digit, and necessarily $d_4 = 1$. It follows that in the base 2 representation for 148, $d_5 = d_6 = 0$. We now have

$$148 - (1 \cdot 2^7 + 1 \cdot 2^4) = 148 - (128 + 16) = 4 = 1 \cdot 2^2 + 0 \cdot 2^1 + 0.$$

Thus $148 = 10010100_2$.

Exercise 6.3. Use the method from the examples above to find the base 2 and base 3 representations of 302.

With these examples behind us, we write a program that takes as input a natural number and returns the digits in its base 2 representation. What do we need our program to do? If you look at our last example, you will see that first and foremost we need to be able to find the largest power of 2 that is less than or equal to a number. We want to write a function that does this, because we need to do this step several times in the program; we first find the largest power of 2 less than or equal to our given number n. Once we have found this largest power 2^k, we need to find the largest power of 2 less than or equal to $n - 2^k$, and so on. Let's look at the following function.

```
def largest_power(number):
    i = 0
    while 2**i<=number:
        i=i+1
    return(i-1)
```

What does this function do? For a given input *number*, it starts by setting $i = 0$ and testing whether $1 = 2^0 \leq number$ is true. If it is, i is incremented to 1, and we check whether $2^1 \leq number$ is true. If it is, we increment i to 2 and check whether $2^2 \leq number$ is true, and so on. At some point, we encounter an i (and it is the smallest such) for which $2^i \leq number$ is false. When this happens, we know the largest power of 2 less than or equal to *number* is the one preceding it, namely 2^{i-1}, and we want the function to return this exponent $i - 1$.

Now that we have this function, how do we use it to generate the digits in the base 2 representation of a given number n? Just as in the example above, we start by figuring out how long the base 2 representation of n is. In the program below, this number is called N and is one more than the exponent associated with the largest power of 2 that is less than or equal to n. Once we know N, we can start finding the digits. In the program, we define an empty list to hold these digits. We have to put N things in this list, and we will do this by going through a loop N times, as j ranges from 0 to $N - 1$.

What should happen the first time through the loop? We should append a 1 to our list as the left-most digit in the binary expansion of n *and* we should update n so that the next times through the loop we are finding digits of $n - 2^{N-1}$.

```
# Takes the base 10 representation of a number and returns the
    base 2 representation

n=input("enter the base 10 representation of a natural number: ")

# finds the largest power of 2 less than or equal to a number
def largest_power(number):
    i = 0
    while 2**i<=number:
```

```
        i=i+1 #as long as 2**i<=number, we want to increment i
            and go through again
    return(i-1) #we don't execute the while statement once
        2**i>number. i-1 is then the largest power <=number.

# make the sequence of binary digits
N=largest_power(n)+1 #number of digits in the binary
    representation of n
binary_digits=[] #define an empty list to hold the binary digits
temp=n
for j in range(0,N):
    if N-1-j==largest_power(temp):
        binary_digits.append(1)
        temp=temp-2**largest_power(temp)
    else:
        binary_digits.append(0)

print(binary_digits)
```

To make sure we understand how the algorithm works, let's see what it does when the input is $n = 73$. We first determine N. Because the largest power function returns 6, $N = 7$. Thus the output of the algorithm will be a list with 7 digits. Let's consider what happens each time through the *for* loop. The first time through, $temp = 73$ and $j = 0$. Because $N - 1 - j = 6$ and $6 = largest\_power(73)$, we append 1 to the list *binary_digits* and replace *temp* with $73 - 2^6 = 9$.

The second time through the loop, $j = 1$ and $temp = 9$. $N - 1 - j = 5$, but $largest\_power(9) = 3$, so we append 0 to *binary_digits* and do not change *temp*. The third time through the loop, $j = 2$, and $N - 1 - j = 4$. Because this is still not equal to $largest\_power(9) = 3$, we again append 0 to *binary_digits* and do not change *temp*. When $j = 3$, $N - 1 - j$ does equal *largest_power*(9) and we append 1 to the list and replace *temp* with $9 - 2^3 = 1$. Because $largest\_power(1) = 0$, when $j = 4$ and $j = 5$, we append 0 when we go through the loop and do not change *temp*. Finally, when $j = 6 = N - 1$, $N - 1 - j = 0$ and we append 1 to *binary_digits*. The entire program returns

$$[1, 0, 0, 1, 0, 0, 1].$$

Exercise 6.4. Trace your age through the program as we did above to see that you obtain the expected result.

6.1.2 Proof of the existence and uniqueness of the base b representation of an element of $\mathbb{N}$

We are now ready to prove our theorem on base b representations of natural numbers. If you carefully studied the algorithm in the previous section, this proof should seem easy.

Theorem 6.1. *Let n be a natural number and let b be a natural number satisfying $b \geq 2$. Then there exist a non-negative integer N and non-negative integers d_j with $0 \leq d_j \leq b-1$ such that*

$$n = \sum_{j=0}^{N} d_j b^j. \tag{6.4}$$

Proof. The proof is by strong induction on n. If $n = 1$, we may write

$$1 = 1 \cdot b^0.$$

Because $2 \leq b$, $1 \leq b-1$, and we indeed have a valid base b representation for $n = 1$.

Now suppose the result holds for all $1 \leq i \leq k$, i.e., that any such i has a base b representation. Consider $k+1$. Find N such that

$$b^N \leq k+1 < b^{N+1}.$$

Because $b^0 = 1$ and the sequence $\{b^j\}$ is unbounded, N exists. We now define d_N to be a non-negative integer such that

$$d_N b^N \leq k+1 < (d_N+1)b^N.$$

Because $k+1 < b^{N+1} = b \cdot b^N$ and $d_N + 1$ is the smallest integer m for which $k+1 < mb^N$, necessarily $d_N + 1 \leq b$, or $d_N \leq b-1$. Thus d_N is a valid digit. Also note that, because $b^N \leq n$ and d_N is the largest integer m for which $mb^N \leq k+1$, $1 \leq d_N$.

Now consider $k+1-d_N b^N$. Call this number k_1. If $k_1 = 0$, $k+1 = d_N b^N$, and we have found our base b representation for b. If $k_1 \neq 0$, then $1 \leq k_1 \leq k$ and so by the induction hypothesis, k_1 has a valid base b representation

$$k_1 = \sum_{j=0}^{M} d'_j b^j$$

for some M with $d_M \geq 1$. Because M is the smallest integer with the property that $k_1 < b^{M+1}$ but also $k_1 < k+1 < b^{N+1}$, $M \leq N$. We claim that in fact $M < N$. Suppose not. If $M = N$, then

$$\begin{aligned} k+1 &= d_N b^N + k_1 \\ &\geq d_N b^N + d'_N b^N \\ &\geq (d_N+1)b^N. \end{aligned}$$

This inequality contradicts our definition of d_N.

We may now write

$$k+1 = d_N b^N + \sum_{j=0}^{M} d'_j b^j.$$

If we set $d_j = 0$ if $M+1 \leq j \leq N-1$ and $d_j = d'_j$ if $0 \leq j \leq M$, we have indeed obtained a valid base b representation for $k+1$. □

We have now shown that every natural number has a base b representation. We would also like to know that such a representation is unique. This result is established by proving the following proposition.

Proposition 6.1. *Let b be a natural number satisfying $b \geq 2$. Let N, M be non-negative integers and let c_j, d_j be non-negative integers satisfying $0 \leq c_j, d_j \leq b-1$. Suppose also $d_N, c_M \neq 0$. Then if*

$$\sum_{j=0}^{N} d_j b^j = \sum_{j=0}^{M} c_j b^j,$$

then $N = M$ and $d_j = c_j$ for all $0 \leq j \leq N$.

Why does the uniqueness of the base b representation of a natural number follow from this proposition? The basic idea is that, when we want to prove that there is a unique object with a certain property, we suppose there are two such objects and prove that, in fact, the two objects must be identical. Thus for a given natural number n, we suppose that $\sum_{j=0}^{N} d_j b^j$ and $\sum_{j=0}^{M} c_j b^j$ are two valid base b representations of n and we show that $d_j = c_j$ for all j.

The next exercise, which is challenging, guides you through the proof of Proposition 6.1.

Exercise 6.5. (Notation as defined in Proposition 6.1.)

(a) Show that

$$(b-1)\sum_{j=0}^{M} b^j = (b-1)(b^M + \ldots + b^2 + b + 1).$$

(Hint: Recall the formula for the sum of a finite geometric series.)

(b) Suppose now, as in Proposition 6.1, that $\sum_{j=0}^{N} d_j b^j = \sum_{j=0}^{M} c_j b^j$. Denote by n the common value of this sum. Prove that

$$b^N \leq n < b^{M+1}$$

and

$$b^M \leq n < b^{N+1}.$$

(c) Prove that $N = M$.

(d) We now know that $N = M$. Prove that

$$d_N b^N \leq n < (c_N + 1)b^N$$

and that

$$c_N b^N \leq n < (d_N + 1)b^N.$$

(e) Conclude that $c_N = d_N$.

(f) Conclude that $d_j = c_j$ for all $0 \leq j \leq N$.

6.2 Integers and Divisibility

We briefly studied divisibility in Chapter 1. We recall the definitions and establish some notation.

Definition 6.2. Let $a, b \in \mathbb{Z}$ with $b \neq 0$. We say b **divides** a and write $b|a$ if there exists an integer m such that $a = mb$. When $b|a$ we say b is a **factor** of a or a **divisor** of a. A natural number p different from 1 is **prime** if its only non-negative factors are 1 and p.

Proposition 6.2. *Let a, b, and c be non-zero integers.*

(a) If $a|b$ and $b|c$, then $a|c$. Thus the relation $|$ is transitive.

(b) If $c|a$ and $c|b$, then $c|(ma+nb)$ for all $m, n \in \mathbb{Z}$.

Proof. We prove part (b) and leave the proof of part (a) for the next exercise.

Because $c|a$ and $c|b$, there exist integers j and k such that $a = jc$ and $b = kc$. For any integers m and n,

$$ma + nb = mjc + nkc = (mj + nk)c,$$

and thus $c|(ma + nb)$. □

Exercise 6.6. Prove part (a) of Proposition 6.2. Is $|$ an equivalence relation on $\mathbb{Z} \setminus \{0\}$?

If a and b are integers, and if $d|a$ and $d|b$, we call d a *common divisor* of a and b. If a and b are not both zero, we call the largest such d the *greatest common divisor* of a and b and denote it by $\gcd(a, b)$.

Definition 6.3. Let $a, b \in \mathbb{Z}$ with a and b not both zero. If $\gcd(a, b) = 1$, we say that a and b are **relatively prime**.

Example 6.3. If $a = 12$ and $b = 16$, $\gcd(a, b) = 4$. On the other hand, if $a = 9$ and $b = 16$, $\gcd(a, b) = 1$. Thus 9 and 16 are relatively prime.

The idea of the greatest common divisor of two integers is probably familiar to you. You find it whenever you want to reduce a fraction to lowest terms. In the example above, because $\gcd(12, 16) = 4$, we may write

$$\frac{12}{16} = \frac{3 \cdot 4}{4 \cdot 4} = \frac{3}{4},$$

and this last fraction is in lowest terms. On the other hand, because $\gcd(9, 16) = 1$, the fraction $\frac{9}{16}$ is already in lowest terms.

Whereas the concept of the greatest common divisor of two numbers is familiar, the next proposition is probably not. It is, however, one of the most useful results about relatively prime integers.

Proposition 6.3. *Let a and b be integers, not both zero. If* $\gcd(a,b) = 1$, *then there exist integers m and n such that*

$$1 = ma + nb.$$

An expression of the form $ma + nb$ for $m, n \in \mathbb{Z}$ is called an *integer combination* of a and b. Before we prove the proposition, we give a few exercises that serve as lemmas.

Exercise 6.7. Let c and d be positive integers. Prove that, if there is an integer combination of c and d equalling 1, there is an integer combination of c and $-d$ equalling 1, and similarly for the pair $-c$ and d or the pair $-c$ and $-d$.

Exercise 6.8. Let c and d be positive integers. Show that

$$\gcd(c,d) = \gcd(-c,d) = \gcd(c,-d) = \gcd(-c,-d).$$

Exercise 6.9. Show that $\gcd(c,0) = 1$ if and only if $c = \pm 1$.

Exercise 6.10. Show that $\gcd(c,c) = 1$ if and only if $c = \pm 1$.

As a consequence of these four exercises, in our proof of Proposition 6.3 we will be able to treat only the case in which a and b are non-negative integers and $a > b$. We require one more lemma about relatively prime integers.

Lemma 6.1. *Let a and b be integers, not both zero, with* $\gcd(a,b) = 1$. *Then* $\gcd(b, a-b) = 1$.

Proof. Let $d|b$ and $d|(a-b)$. We claim $d|a$. Indeed, because $d|b$, there exists $j \in \mathbb{Z}$ such that $b = jd$. Because $d|(a-b)$, there exists $k \in \mathbb{Z}$ such that $a-b = kd$. Thus

$$a = b + kd = jd + kd = (j+k)d,$$

and $d|a$. As a consequence, $\gcd(b, a-b)$ is a divisor of both a and b. Because the greatest common divisor of a and b is 1, $\gcd(b, a-b)$ can not exceed 1. Because 1 is always a common divisor of any pair of numbers, we conclude $\gcd(b, a-b) = 1$. $\square$

We turn now to the proof of Proposition 6.3.

Proof. By the exercises above, it suffices to prove the result for non-negative integers a, b with $a > b$. The proof will be by induction on $a + b$.

When $a+b = 1$, the result holds with $m = n = 1$. Thus we assume the result holds when the sum of the integers is between 1 and k for some natural number k and we consider the case for $k+1$. Thus we suppose we have relatively prime non-negative integers a and b with $a > b$ and $a + b = k + 1$. Now, if $b = 0$, $\gcd(a,b)$ can only be 1 if $a = 1$. Because $a + b = k + 1 \geq 2$, this is not the case. Thus we know $b > 0$. Consider the integers b and $a - b$. By Lemma 6.1, $\gcd(b, a-b) = 1$. Furthermore, because b and $a - b$ are both positive and

$b+(a-b) = a < a+b = k+1$, by the inductive hypothesis there exist integers p and q such that

$$pb + q(a-b) = 1.$$

Rearranging, we find

$$qa + (p-q)b = 1,$$

which is an integer combination of a and b equalling 1. Thus the result holds for $a+b = k+1$. By the principle of mathematical induction, it holds when $a+b = k$ for all natural numbers k. □

We explore several consequences of Proposition 6.3.

Proposition 6.4. *Let a and b be relatively prime and suppose $a|cb$. Then $a|c$.*

Proof. Because a and b are relatively prime, there exist integers m and n so that $1 = ma + nb$. Thus $c = cma + cnb$. Because $a|cb$, there exists k such that $cb = ka$. Thus

$$c = cma + nka = (cm + nk)a,$$

that is, $c = da$ for the integer $d = cm + nk$. Therefore $a|c$. □

The next proposition may seem simple, but it is one of the most commonly used facts about prime numbers.

Proposition 6.5. *Let p be a prime and suppose $p|ab$. Then either $p|a$ or $p|b$.*

Proof. Suppose $p|ab$ but that p does not divide a. Then $\gcd(p, a) = 1$. By Proposition 6.4, $p|b$. □

With this background, we want to solve a practical problem: Given two (possibly very large) integers a and b, how can we find their greatest common divisor without factoring them? The *Euclidean algorithm* provides an answer. We will assume for the remainder of the section that a and b are positive integers, for we know that $\gcd(a, b) = \gcd(|a|, |b|)$. Before describing the algorithm, we establish several useful lemmas.

Lemma 6.2. *(Division algorithm) Let a and b be positive integers. Then there exist unique non-negative integers q and r such that*

$$a = qb + r \quad \textit{and} \quad 0 \leq r < b. \tag{6.5}$$

Proof. First we prove the existence of a representation (6.5). If $b = 1$, then $a = a \cdot 1 + 0$, which is a representation of the form (6.5). Thus for the remainder of the proof we may think of b as a fixed natural number greater than 1.

The proof is by induction on a. When $a = 1$, $1 = 0 \cdot b + 1$ is a valid representation of the form (6.5), and the result holds.

Now suppose the result holds for a, so that there exist q, r with $0 \leq r < b$ such that $a = qb + r$. Now consider $a + 1$. Clearly

$$a + 1 = qb + r + 1.$$

If $r < b-1$, then $r+1 < b$ and we have achieved our desired representation for $a+1$. If, on the other hand, $r = b-1$, then $r+1 = b$ and

$$a+1 = qb + b = (q+1)b + 0$$

is a valid representation for $a+1$. We have now shown that the result holds for $a+1$. By the principle of mathematical induction, it holds for all a.

We now prove the uniqueness statement. Thus we suppose a is a positive integer and that we have integers q_1, q_2, r_1, and r_2 with $0 \le r_1, r_2 < b$ such that $a = q_1 b + r_1$ and $a = q_2 b + r_2$. Then

$$r_1 - r_2 = (q_2 - q_1)b.$$

Thus $b|(r_1 - r_2)$. On the other hand, $0 \le r_1 < b$ and $-b < -r_2 \le 0$ imply

$$-b < r_1 - r_2 < b.$$

Thus $r_1 - r_2 = 0$, i.e., $r_1 = r_2$. It follows that $(q_2 - q_1)b = 0$ and so $q_2 - q_1 = 0$ and $q_1 = q_2$. The uniqueness of the representation (6.5) is established. □

The idea of the Euclidean algorithm is that, if we want to compute $\gcd(a,b)$ for $a > b$, we first use the division algorithm to write $a = qb + r$ for $0 \le r < b$ and then compute $\gcd(b,r) = \gcd(b, a - qb)$, which is presumably much easier because the integers involved are smaller. This approach is legitimate in light of the following lemma.

Lemma 6.3. *Let a and b be integers, not both zero. Then for any integer q,*

$$\gcd(a,b) = \gcd(b, a - qb).$$

Proof. Let S be the set of all common divisors of a and b and let T be the set of all common divisors of $a - qb$ and b. We prove $S = T$ by proving two containments.

Suppose first that $d \in S$. Then there exist $j, k \in \mathbb{Z}$ such that $a = jd$ and $b = kd$. Then

$$a - qb = jd - qkd = (j - qk)d,$$

and so $d|(a - qb)$. It follows that $d \in T$, and so $S \subseteq T$.

For the reverse containment, suppose that $d \in T$. Then there exist $j, k \in \mathbb{Z}$ such that $a - qb = jd$ and $b = kd$. Then

$$a = qb + jd = qkd + jd = (qk + j)d,$$

i.e., $d|a$. Thus $d \in S$ and $T \subseteq S$.

We conclude that $S = T$ and that the common divisors of a and b are the same as the common divisors of $a - qb$ and b. Thus $\gcd(a,b) = \gcd(b, a - qb)$. □

We turn now to the Euclidean algorithm. The algorithm takes as input positive integers a and b with $a > b$. It then computes the remainder r when a is divided by b. By Lemma 6.3, $\gcd(a,b) = \gcd(b,r)$. If $r = 0$, $\gcd(b,r) = b$ and we are done. Else we repeat the process with the pair (b,r). The process terminates when one member of the pair is zero.

Before we look at the code, let us consider some examples.

Example 6.4. We find $\gcd(574, 198)$. Because

$$574 = 2(198) + 178,$$

$\gcd(574, 198) = \gcd(198, 178)$. Because

$$198 = 1(178) + 20,$$

$\gcd(198, 178) = \gcd(178, 20)$. Because

$$178 = 8(20) + 18,$$

$\gcd(178, 20) = \gcd(20, 18)$. Because

$$20 = 1(18) + 2,$$

$\gcd(20, 18) = \gcd(18, 2)$. Finally, because

$$18 = 9(2) + 0,$$

$\gcd(18, 2) = \gcd(2, 0) = 2$. Thus $\gcd(574, 198) = 2$.

Exercise 6.11. Use the Euclidean algorithm to find $\gcd(768, 424)$.

The code for this algorithm is quite simple.

```
# Euclidean algorithm for finding the gcd of two positive
    integers a and b with a > b.

def greatest_common_divisor(a,b):
    larger=a #initializes larger and smaller as a and b
    smaller=b
    while smaller > 0:
        r=larger%smaller #finds the remainder when larger is
            divided by smaller
        larger=smaller #replaces larger with smaller
        smaller=r #replaces smaller with the remainer r
    return(larger)
```

We must prove that this algorithm works.

Theorem 6.2. *Let a, b be positive integers. Then the Euclidean algorithm returns* $\gcd(a, b)$.

Proof. Let (a_n, b_n) be the pair of integers (larger,smaller) after the indented code in the *while* loop has been executed for the n-th time. Thus (a_0, b_0) is the initial pair of positive integers (a, b).

First, we claim that, for any n, $\gcd(a_n, b_n) = \gcd(a, b)$. The proof is by induction on n. When $n = 0$, the result is immediate because $(a_0, b_0) = (a, b)$. For the inductive step, suppose $\gcd(a_k, b_k) = \gcd(a, b)$ and consider the case for $k+1$. If $b_k = 0$, we exit the loop at the conditional statement when we encounter it for the $(k+1)$-st time and the pair (a_k, b_k) is unchanged. Otherwise, if $b_k > 0$, we set a_{k+1} equal to b_k and b_{k+1} equal to r where r is the remainder when a_k is divided by b_k. By Lemma 6.3, $\gcd(a_{k+1}, b_{k+1}) = \gcd(a_k, b_k)$, which is $\gcd(a, b)$ by the inductive hypothesis. Thus the result holds for $k+1$, and, by the principle of mathematical induction, it holds for all n.

Next, we must argue that the process in the algorithm terminates, i.e., that we only go through the loop a finite number of times. Observe that, because b_{k+1} is the remainder when a_k is divided by b_k, $b_k > b_{k+1} \geq 0$ for all k. Because $b = b_0$, $b - k \geq b_k \geq 0$. Thus the algorithm must terminate in at most b steps.

Finally, we must know that when it terminates, it returns the correct value. Thus suppose K is the natural number such that $b_K = 0$. Then $\gcd(a, b) = \gcd(a_K, b_K) = \gcd(a_K, 0) = a_K$. Because a_K is the value of *larger* after step K, the algorithm indeed returns a value equal to $\gcd(a, b)$. □

If you go through the steps of the Euclidean algorithm, at step k you may write b_k in terms of a_{k-1} and b_{k-1}. We obtain a method for writing $\gcd(a, b)$ as an integer combination of a and b. For example, consider $a = 20$ and $b = 12$. Then $20 = 1(12) + 8$. Thus $a_1 = 12$, $b_1 = 8$, and

$$8 = 20 - 12 = a - b.$$

Now we consider the pair $a_1 = 12$, $b_1 = 8$. Then $12 = 1(8) + 4$. Thus $a_2 = 8$, $b_2 = 4$, and

$$4 = 12 - 8 = b - (a - b) = -a + 2b.$$

Observe that $4 = \gcd(20, 12)$ and that this last equation expresses 4 as an integer combination of 20 and 12.

Exercise 6.12. Express $\gcd(100, 64)$ as an integer combination of 100 and 64.

6.3 Modular Arithmetic

6.3.1 Definition of congruence and basic properties

We begin this section with an elementary example.

Example 6.5. Consider a standard twelve-hour clock. Suppose it currently reads 10:00. What will it read in 3 hours? In 17 hours? In 115 hours?

We know without doing any calculation that in 3 hours it will read 1:00. If we wanted to think of how to do this as an addition problem, we could find 10+3=13. We know that 13 is never displayed on a twelve-hour clock but is equivalent to 1. What about after 17 hours? Clearly $10 + 17 = 27$. To what number on a clock is 27 equivalent? Because 24 is equivalent to 12, 27 is equivalent to 3. Written another way, $27 = (2)(12) + 3$. Finally, what about after 115 hours? Because $10 + 115 = 125 = (10)(12) + 5$, the clock will read 5:00.

In this example, we do not care about the sum of the integers but rather about its remainder when divided by 12. We are doing *modular arithmetic.* In this section, we define modular arithmetic carefully and explore some elementary properties. We will delve even deeper later in the chapter when we discuss various algebraic structures.

In Example 4.6, we defined an equivalence relation on the integers by the rule $n \sim m$ if 3 divides $n - m$. We generalize this notion.

Definition 6.4. Let d be a natural number with $d \geq 2$. Define a relation $\equiv$ on $\mathbb{Z}$ by $n \equiv m \pmod{d}$ if d divides $n - m$. We call this relation **congruence modulo** d.

A simple modification of the argument in Example 4.6 establishes that congruence modulo d is an equivalence relation.

Exercise 6.13. Let d be a natural number with $d \geq 2$. Show that the relation defined in Definition 6.4 is an equivalence relation.

Example 6.6. Returning to the clock example above, we can write the results succinctly using our new concept. We are working modulo 12. $10 + 3 \equiv 1 \pmod{12}$ because 12 divides $13 - 1$. $10 + 17 \equiv 3 \pmod{12}$ because 12 divides $27 - 3$. Finally, $10 + 115 \equiv 5 \pmod{12}$ because 12 divides $125 - 5$.

We establish some elementary properties of congruence.

Proposition 6.6. *Let d be a natural number with $d \geq 2$. Let $a, A, b, B \in \mathbb{Z}$ with $a \equiv A \pmod{d}$ and $b \equiv B \pmod{d}$. Then*

(a) $a + b \equiv A + B \pmod{d}$

(b) $ab \equiv AB \pmod{d}$.

Proof. We prove part (b). The proof of part (a) is easier and is left as an exercise.

Because $a \equiv A \pmod{d}$, there exists $j \in \mathbb{Z}$ such that $a - A = jd$. Similarly, there exists $k \in \mathbb{Z}$ such that $b - B = kd$. We consider $ab - AB$. We would like to rewrite this expression in terms of $a - A$ and $b - B$. We begin by adding a clever form of 0. Then we factor.

$$\begin{aligned} ab - AB &= ab - Ab + Ab - AB \\ &= (a - A)b + A(b - B) \\ &= jdb + Akd \\ &= (jb + kA)d. \end{aligned}$$

Thus $d|(ab - AB)$ and indeed $ab \equiv AB \pmod{d}$. □

Exercise 6.14. Prove part (b) of Proposition 6.6

6.3.2 Congruence classes

We explore in more detail congruence classes modulo d. First we recall our notation:

$$\bar{a} = \{b \in \mathbb{Z} \mid b \equiv a \pmod{d}\}.$$

How many congruence classes modulo d are there? By the Division Algorithm (Lemma 6.2), for any integer a there is a unique integer r with $0 \leq r \leq d-1$ such that $a = qd+r$, or, equivalently, $a \equiv r \pmod{d}$. Furthermore, the numbers $0, 1, \ldots, d-1$ are all in distinct congruence classes. Indeed, if $0 \leq r, s \leq d-1$,

$$-(d-1) \leq r - s \leq d-1.$$

Thus we may only have $r \equiv s \pmod{d}$ if $r = s$. We conclude that there are precisely d congruence classes modulo d. We often use the d integers $0, 1, \ldots, d-1$ as our preferred representatives and use $\bar{0}, \bar{1}, \ldots, \overline{d-1}$ as our preferred labels for the classes.

Often our goal is to determine the congruence class containing the result of an operation with integers. For such problems, Proposition 6.6 is a useful tool.

Example 6.7. Suppose we want to determine the congruence class of $43 \cdot 26$ modulo 7. Of course one approach is to multiply 43 and 26 and then find the remainder when the result is divided by 7. Another approach is to observe that $43 \equiv 1 \pmod{7}$ and $26 \equiv 5 \pmod{7}$. Then by Proposition 6.6,

$$43 \cdot 26 \equiv 1 \cdot 5 \pmod{7}.$$

Thus $43 \cdot 26$ is in the class $\bar{5}$.

Exercise 6.15. Find the congruence class of $65 \cdot 84$ modulo 9.

Exercise 6.16. Find the congruence class of $1243 \cdot 7268$ modulo 10. Then describe a simple way of quickly determining the congruence class of nm modulo 10 for any integers n and m.

We have already observed that the d consecutive integers $0, 1, \ldots, d-1$ are in distinct equivalence classes. The next proposition is a generalization of this observation.

Proposition 6.7. *Consider any set S of d consecutive integers. Then the elements of S represent all d distinct congruence classes modulo d.*

Proof. Denote the smallest element of S by n. Then $S = \{n+j : 0 \leq j \leq d-1\}$. Suppose $n+j$ and $n+k$ are elements of S in the same congruence class modulo d. Then there exists an integer l such that $(n+j)-(n+k) = ld$, or, equivalently, $j - k = ld$. Because $0 \leq j, k \leq d-1$,

$$-(d-1) \leq j - k \leq d-1.$$

Thus $l = 0$ and $j = k$. Thus distinct elements of S are in distinct congruence classes modulo d. Because there are d elements of S and d congruence classes modulo d, the result follows. □

Proposition 6.7 allows us to prove some interesting divisibility results.

Proposition 6.8. *Any product of d consecutive integers is divisible by d.*

Proof. By Proposition 6.7, the d consecutive integers include precisely one element from each congruence class modulo d. Thus the product is congruent modulo d to $0 \cdot 1 \cdot \ldots \cdot (d-1) = 0$. In other words, the product is divisible by d. □

We can now give another proof that, for any integer n, $n^2 - n$ is even. Indeed, $n^2 - n = n(n-1)$ is a product of two consecutive integers, and so by Proposition 6.8, it is divisible by 2. The problems at the end of the chapter include some more applications of this idea.

6.3.3 Operations on congruence classes

Proposition 6.6 tells us that when we do addition and multiplication of integers, the congruence class of the result depends only on the congruence classes of the integers on which we operate. This fact allows us to now define operations *on the congruence classes themselves.*

For the development that follows, we fix a natural number $d \geq 2$. We denote by $\mathbb{Z}_d$ the set of congruence classes modulo d.

Definition 6.5. Let $\bar{a}, \bar{b} \in \mathbb{Z}_d$. Let A be any element of $\bar{a}$ and let B be any element of $\bar{b}$. We define $\bar{a} + \bar{b}$ to be the congruence class of $A + B$ and we define $\bar{a} \cdot \bar{b}$ to be the congruence class of AB.

This definition takes some getting used to, so we illustrate with some examples.

Example 6.8. Let $d = 5$. Suppose we want to find $\bar{1} + \bar{4}$. The definition says we take any elements we want out of $\bar{1}$ and $\bar{4}$, add them, and report the congruence class of the result. For example, we can take 1 from $\bar{1}$ and 4 from $\bar{4}$. Their sum is $1 + 4 = 5$, which is in $\bar{0}$. We could choose different elements from the congruence classes. For example, we could take 6 from $\bar{1}$ and 14 from $\bar{4}$. Their sum is $6 + 14 = 20$, which is of course also in $\bar{0}$. We summarize our calculation in either case by writing $\bar{1} + \bar{4} = \bar{0}$.

Exercise 6.17. Do the following calculations in $\mathbb{Z}_6$.

(a) $\bar{2} + \bar{5}$.

(b) $\bar{2} \cdot \bar{5}$.

(c) $\bar{5} \cdot \bar{5}$.

The procedure we have outlined for arithmetic with congruence classes may seem somewhat tortured, but it is a very common way of defining operations on sets of equivalence classes. We will do something similar when we talk about operations on rational numbers later in this chapter and when we talk about the real number system later in this book. In all of these situations, we will have something important to prove. *We must prove that when we define an operation on equivalence classes in terms of a known operation on representatives from those classes, the equivalence class of the answer is independent of our choice of representative.* When such is the case, we will say that the operation under consideration is *well-defined.* As we already remarked at the beginning of this subsection, by Proposition 6.6, our operations in $\mathbb{Z}_d$ are well-defined.

Our operations satisfy a number of familiar and useful properties.

Proposition 6.9. *Let* $\bar{a}, \bar{b}, \bar{c} \in \mathbb{Z}_d$.

(a) (associativity of operations) $\bar{a}+(\bar{b}+\bar{c}) = (\bar{a}+\bar{b})+\bar{c}$ *and* $\bar{a}\cdot(\bar{b}\cdot\bar{c}) = (\bar{a}\cdot\bar{b})\cdot\bar{c}$.

(b) (commutativity of operations) $\bar{a}+\bar{b} = \bar{b}+\bar{a}$ *and* $\bar{a}\cdot\bar{b} = \bar{b}\cdot\bar{a}$.

(c) (existence of additive and multiplicative identities) $\bar{a}+\bar{0} = \bar{a}$ *and* $\bar{a}\cdot\bar{1} = \bar{a}$.

(d) (existence of additive inverses) $\bar{a}+\overline{d-a} = \bar{0}$.

The proof is easy and we leave most of it to the exercises. It mostly uses corresponding properties for the operations with integers. Our main reason for stating the proposition is that later in the chapter we will discuss various sets of axioms a set and its operations may satisfy, and we will want to use $\mathbb{Z}_d$ as an example.

Proof. We prove commutativity of addition. Take $a \in \bar{a}$ and $b \in \bar{b}$. By definition, $\bar{a}+\bar{b}$ is the congruence class containing $a+b$ and $\bar{b}+\bar{a}$ is the congruence class containing $b+a$. Because a and b are integers and addition of integers is commutative, $a+b = b+a$. Thus the class containing $a+b$ is the same as the class containing $b+a$, i.e., $\bar{a}+\bar{b} = \bar{b}+\bar{a}$. □

Exercise 6.18. Prove the remaining parts of Proposition 6.9.

We end this section with a less trivial result about multiplication.

Theorem 6.3. *If p is prime, then every non-zero element of $\mathbb{Z}_p$ has a multiplicative inverse.*

Proof. Let $\bar{a}$ be a non-zero element of $\mathbb{Z}_p$. To show that $\bar{a}$ has a multiplicative inverse in $\mathbb{Z}_p$, we must show that for some $0 \le j \le p-1$, $\bar{j}\bar{a} = \bar{1}$.

Consider the p integers ja for $0 \le j \le p-1$. We claim that no two of these integers are in the same congruence class. Indeed, if $ja \equiv ka \pmod{p}$ for some j and k with $0 \le j, k \le p-1$, then $p|(j-k)a$. Because p is prime, by Proposition 6.5 either $p|(j-k)$ or $p|a$. The latter is impossible, for if $p|a$, $\bar{a} = \bar{0}$. Thus $p|(j-k)$. Because $-(p-1) \le j-k \le p-1$, we conclude that $j-k = 0$. Because the p integers ja, $0 \le j \le p-1$, are in different congruence classes and there are a total of exactly p congruence classes, there exists J so that $Ja \equiv 1 \pmod{p}$. In other words, $\bar{J}\bar{a} = \bar{1}$. We have found the multiplicative inverse of $\bar{a}$. □

The above proof shows that all non-zero elements of $\mathbb{Z}_p$ have multiplicative inverses if p is prime, but it does not tell how to find multiplicative inverses.

Exercise 6.19. Find the multiplicative inverses of all non-zero elements of $\mathbb{Z}_7$.

6.4 The Rational Numbers

We think of a rational number as a ratio $\frac{p}{q}$ where p and q are integers with $q \neq 0$. A moment's thought indicates that there is a little more to it than that, for although $\frac{1}{2}$, $\frac{2}{4}$, and $\frac{50}{100}$ are not the same ratios of integers, we think of them as representing the same rational number. Our purpose in this section is to use the ideas we have developed so far in this book to give a definition of rational numbers that incorporates into it the notion that a given rational number can be represented by more than one ratio of integers.

Consider the set $S = \mathbb{Z} \times (\mathbb{Z} \setminus \{0\})$ and denote elements of S by $\frac{a}{b}$ instead of (a, b). Define a relation $\sim$ on S by

$$\frac{a}{b} \sim \frac{A}{B} \quad \Leftrightarrow \quad aB = Ab. \tag{6.6}$$

Proposition 6.10. *The relation* $\sim$ *defined in* (6.6) *is an equivalence relation on* S.

Proof. We must show that $\sim$ is reflexive, symmetric, and transitive. The first two are clear, for if $\frac{a}{b} \in S$, $ab = ab$ and $\frac{a}{b} \sim \frac{a}{b}$. Also, if $\frac{a}{b} \sim \frac{A}{B}$, $aB = Ab$. Because then $Ab = aB$, $\frac{A}{B} \sim \frac{a}{b}$.

Finally, suppose $\frac{a}{b} \sim \frac{A}{B}$ and $\frac{A}{B} \sim \frac{\alpha}{\beta}$. Then $aB = Ab$ and $A\beta = \alpha B$. Then

$$a\beta = \left(\frac{Ab}{B}\right)\beta = \frac{b}{B}(A\beta) = \frac{b}{B}(\alpha B) = \alpha b,$$

i.e., $\frac{a}{b} \sim \frac{\alpha}{\beta}$. □

Because $\sim$ is an equivalence relation on S, it partitions S into disjoint equivalence classes. We are ready for the definition of the rational numbers.

Definition 6.6. The set $\mathbb{Q}$ of **rational numbers** is the set of all equivalence classes of $\mathbb{Z} \times (\mathbb{Z} \setminus \{0\})$ under the equivalence relation defined in (6.6).

Although the preceding definition takes a bit of getting used to, it really does coincide with the way you have learned to think about rational numbers. When you think about the rational number "one quarter," part of your understanding of that number includes the fact that it can be represented as a ratio in multiple way. The ratios $\frac{1}{4}$, $\frac{25}{100}$, $\frac{-1}{-4}$, and $\frac{4}{16}$ (to name but a few) are all valid ways to represent the rational number "one quarter." Each one of these ratios is in the equivalence class that we could call "one quarter." Each of them is a different representative for that equivalence class. Different representatives may be appropriate for different situations.

Exercise 6.20. You think of $\frac{1}{4}$ and $\frac{25}{100}$ as equivalent fractions. Show that in fact $\frac{1}{4} \sim \frac{25}{100}$, with $\sim$ as defined in (6.6). Then describe all other ratios $\frac{a}{b}$ with $b \neq 0$ such that $\frac{1}{4} \sim \frac{a}{b}$.

You are, of course, very comfortable doing arithmetic with rational numbers. In grade school we all learn procedures for adding, subtracting, multiplying, and dividing fractions. We also believe that we may replace any fraction at any time with another equivalent fraction without changing the rational number that is the result. For example, we would consider both of the following calculations to be correct.

$$\frac{1}{6}+\frac{1}{4}=\frac{2}{12}+\frac{3}{12}=\frac{5}{12} \tag{6.7}$$

$$\frac{1}{6}+\frac{1}{4}=\frac{4}{24}+\frac{6}{24}=\frac{10}{24}=\frac{5}{12}. \tag{6.8}$$

Let's look more closely at arithmetic operations on rational numbers in light of our new understanding of rational numbers as equivalence classes. Suppose we want to add two rational numbers r and s. Because rational numbers are equivalence classes of fractions, we could take a fraction $\frac{a}{b}$ from the equivalence class r and a fraction $\frac{c}{d}$ from the equivalence class s. These fractions are now just the sort of thing you worked with in grade school. Combine them as you did then, by getting a common denominator:

$$\frac{a}{b}+\frac{c}{d}=\frac{ad}{bd}+\frac{bc}{bd}=\frac{ad+bc}{bd}. \tag{6.9}$$

The result of this calculation is another fraction, that is, an element of S. We want the result of the original addition problem to be a rational number, or, in other words, an equivalence class of fractions. We therefore define the sum of the rational numbers to be the equivalence class containing $\frac{ad+bc}{bd}$.

We state this idea more concisely in the following definition.

Definition 6.7. Let $r, s \in \mathbb{Q}$. Let $\frac{a}{b}$ be a representative of r and $\frac{c}{d}$ a representative of s. Then $r+s$ is the equivalence class containing $\frac{ad+bc}{bd}$.

In a similar manner we can define multiplication of rational numbers.

Definition 6.8. Let $r, s \in \mathbb{Q}$. Let $\frac{a}{b}$ be a representative of r and let $\frac{c}{d}$ be a representative of s. Define $r \cdot s$ (or just rs) to be the equivalence class containing $\frac{ac}{bd}$.

Certainly your experience adding fractions leads you to think that these operations are in fact well-defined. Think again about the two calculations done in equations (6.7) and (6.8). In both, we are trying to find the sum of the two rational numbers represented by $\frac{1}{6}$ and $\frac{1}{4}$, respectively. In both solutions, we replace each fraction with a different fraction *in the same equivalence class.* Our aim is to solve the easier arithmetic problem of adding two fractions with the same denominator. We look at the fraction we obtain. Sometimes we report this fraction as our final answer, but sometimes we report a different element of the same equivalence class. Usually we choose as our representative of that equivalence class the fraction that is in "lowest terms." We believe that the result of the calculation is independent of the choices of representatives we make at various points in the process.

Of course we know at this point in the course that believing that our operations are well-defined is not enough. We require proof.

Theorem 6.4. *The operations of addition and multiplication in $\mathbb{Q}$ described in Definitions 6.7 and 6.8 are well-defined.*

Proof. We give the proof for addition; the proof for multiplication is similar but a bit easier and is left as the next exercise.

Let $r, s \in \mathbb{Q}$. We must show that our definition of $r + s$ is independent of the representatives we choose. Thus let $\frac{a}{b}$ and $\frac{A}{B}$ be any representatives of r, and let $\frac{c}{d}$ and $\frac{C}{D}$ be any representatives of s. According to Definition 6.7, one definition of $r + s$ is as the equivalence class containing $\frac{ad+bc}{bd}$. Another is as the equivalence class containing $\frac{AD+BC}{BD}$. We must show that these equivalence classes are in fact the same. Thus we must show $\frac{ad+bc}{bd} \sim \frac{AD+BC}{BD}$.

Because $\frac{a}{b} \sim \frac{A}{B}$ and $\frac{c}{d} \sim \frac{C}{D}$, $aB = Ab$ and $cD = Cd$. Observe,

$$\begin{aligned}(ad + bc)BD &= aBdD + bBcD \\ &= AbdD + bBCd \\ &= (AD + BC)bd.\end{aligned}$$

Thus $\frac{ad+bc}{bd} \sim \frac{AD+BC}{BD}$ and the proof is complete. □

Exercise 6.21. Show that multiplication of rational numbers, as defined in Definition 6.8, is well-defined.

In Chapter 1, we said that we would accept as known that there is a number system $\mathbb{Q}$ with operations of addition and multiplication satisfying the ordered field axioms. If we wanted to, we could now prove some of these assertions, using as our starting points the properties of the integers and our definitions of operations. For example, the field axioms state that there should exist an element of $\mathbb{Q}$, which we call 1, such that $1r = r1 = r$ for all $r \in \mathbb{Q}$. What is this element 1? We claim it is the equivalence class containing the fraction $\frac{1}{1}$. To show that 1 in fact has the desired property, let r be an arbitrary element of $\mathbb{Q}$ and let $\frac{a}{b}$ be any representative of r. Then $1r$ is, by definition, the equivalence class containing $\frac{1 \cdot a}{1 \cdot b}$. But $1 \cdot a$ and $1 \cdot b$ are now operations *in the integers.* If we assume all the properties of the integers from Chapter 1 hold, $1 \cdot a = a$, $1 \cdot b = b$, and so the product really is the equivalence class containing $\frac{a}{b}$, namely r. A similar (tedious) little argument shows that $r1 = r$. Other field axioms can be established in a similar manner from properties of $\mathbb{Z}$. We suggest verifying two more as exercises, but then we move on.

Exercise 6.22. Every field has an additive identity. What is the additive identity in $\mathbb{Q}$? We will, of course, call it 0. Verify that your proposed additive identity has the property $r + 0 = 0 + r = r$ for all r in $\mathbb{Q}$.

Exercise 6.23. Every non-zero element of a field must have a multiplicative inverse. If r is a non-zero element of $\mathbb{Q}$, describe its multiplicative inverse. Call it r^{-1}. Verify that your proposed multiplicative inverse has the property $rr^{-1} = r^{-1}r = 1$.

The foregoing discussion may seem pedantic. Do mathematicians really think of rational numbers in this way? Yes and no. We know that rational numbers are defined in this manner, as equivalence classes, but the results above show that we can usually ignore this technicality and do operations with representative elements, secure in the knowledge that all of the apparent shenanigans we go through when we get common denominators and reduce fractions can be fully justified. At this point, then, we will return to our previous happy state of equilibrium in which we accept that $\mathbb{Q}$ is indeed an ordered field.

We use the ordered field axioms to establish a proposition we will need later.

Proposition 6.11. *Suppose $r, s \in \mathbb{Q}$ and $r, s > 0$. Then $r^2 < s^2$ if and only if $r < s$.*

Proof. Suppose $0 < r < s$. Because $\mathbb{Q}$ is an ordered field,

$$r^2 < rs < s^2.$$

Conversely, suppose $r, s > 0$ and $r^2 < s^2$. If $r \geq s$, as above we would have $r^2 \geq s^2$. Thus $r < s$. □

6.5 Algebraic Structures

If you are a math major, you will someday take a course called Algebraic Structures or Abstract Algebra. When I first saw a course with such a title listed as an upper-division course in my college catalogue, I remember being puzzled. I though I had mastered algebra pretty well in high school. What I did not realize at that time is that Algebra is an entire branch of mathematics and an active research area today. What, if any, connection is there with high school algebra?

High school algebra is essentially a collection of techniques for solving equations when the variable represents a real number and the operations involved are the standard arithmetic operations on real numbers. It starts with easy things like linear equations and progresses to include more general polynomial, rational, and radical equations. All of the rules for manipulating algebraic equations rely on the properties of the operations of addition and multiplication in the real number system. We made this point back in Chapter 1 when we showed how solving a simple linear equation can be thought of as a systematic application of the field axioms.

The field of Algebra seeks to carry these ideas into other settings. What if instead of working in the rational number system or the real number system, we work in some other set with operations that satisfy the field axioms? Can we still solve equations in the same way? What if our set and operations satisfy some but not all of the properties of a field? What if, for example, not all of our non-zero elements have multiplicative inverses? What if we only have one operation defined on our set? What if the commutative property does not hold? These are not just idle questions. For example, we have already seen that $\mathbb{Z}$

with addition and multiplication is an example of a number system in which most non-zero elements do not have a multiplicative inverse. The set of $n \times n$ matrices with the operations of addition and multiplication satisfies most of the same properties as $\mathbb{Z}$, but the multiplication is not commutative. For yet another example, we can consider the set S_n of all bijections from $\{1, 2, \ldots, n\}$ to $\{1, 2 \ldots, n\}$ with the single operation of function composition. S_n has a surprising number of interesting properties; the operation is associative, there is an identity element for the operation, and every element has an inverse. The point of Algebra is this: rather than study each example as a separate entity, study the lists of properties themselves. See what theorems follow from these axioms alone. These theorems will then apply to any specific example that comes along satisfying those axioms.

You have probably already had a taste of this approach in your mathematical career. If you have taken linear algebra, you are familiar with the notion of a vector space. We define a vector space in terms of a set, an accompanying ground field, and operations of addition and scalar multiplication that must satisfy a list of axioms. We prove theorems about all vector spaces. We then discuss specific examples to which these theorems can be applied, like $\mathbb{R}^n$, vector spaces of polynomials, or vector spaces of functions.

The goal of this section is to give you just the briefest of introductions to the variety of algebraic structures that you will study in detail later. We focus mostly here on routine verification proofs.

6.5.1 Binary operations

Definition 6.9. Let S be a set. A **binary operation** on S is a function from $S \times S$ to S.

Said simply, a binary operation takes two elements of S and produces another element of S. The fact that the result of the operation is again in S is often referred to as the *closure property*. In Chapter 1, we discussed what it means to say that the integers are closed under addition. We are now extending this idea to a broader class of sets and operations.

Example 6.9. Let S be the set $\mathbb{Z}$ of integers and let the binary operation be addition, denoted as usual by $+$. Thus if (n, m) is any pair of integers, the result of the binary operation is the new integer $n + m$.

Example 6.10. Let M_2 be the set of all 2×2 matrices and let the binary operation be matrix multiplication, denoted by juxtaposition. Thus if $A, B \in M_2$, the result of the binary operation is the new 2×2 matrix AB.

Example 6.11. Let S be the set of polynomials of degree n in one variable with real coefficients and consider the operation of polynomial addition. This operation is not a binary operation because the result need not be in S. For example, if $n = 2$, $p(x) = x^2 + x$, and $q(x) = -x^2 + x$, then $(p + q)(x) = 2x$. This sum is not in S because its degree is 1, not 2. If we want to study sets of polynomials, we might instead consider the set P_n of polynomials of degree *less than or equal to* n with the binary operation of polynomial addition.

The next few exercises are in the same spirit as the examples. We describe a set and an operation and ask if it is a binary operation. As the above examples show, the question is usually whether or not the result of the operation is again an element of the set.

Exercise 6.24. Let $\mathbb{R}^2$ be the set of all ordered pairs of real numbers. Determine whether each operation is a binary operation.

(a) Let $v, w \in \mathbb{R}^2$ with $v = (a, b)$ and $w = (c, d)$. Define an operation $+$ by $v + w = (a + c, b + d)$.

(b) Let $v, w \in \mathbb{R}^2$ with $v = (a, b)$ and $w = (c, d)$. Define an operation $\cdot$ by $v \cdot w = ac + bd$.

(c) Let $v, w \in \mathbb{R}^2$ with $v = (a, b)$ and $w = (c, d)$. Define an operation $*$ by $v * w = (ac - bd, ad + bc)$.

(d) Let $v \in \mathbb{R}^2$ with $v = (a, b)$ and let $c \in \mathbb{R}$. Define an operation by the rule $cv = (ca, cb)$.

(e) Each of these is a common operation in $\mathbb{R}^2$. Do you recognize them?

Exercise 6.25. Let n be a natural number and let S_n denote the set of all bijections from $\{1, 2, \ldots, n\}$ to $\{1, 2, \ldots, n\}$. Explain why composition of functions, $\circ$, is a binary operation on S_n.

6.5.2 Groups

Having a binary operation on a set doesn't tell us much. We want to know that the operation satisfies certain useful properties. We obtain different algebraic structures by considering different lists of properties our operation(s) may satisfy. The simplest algebraic structure is that of a *group*.

Definition 6.10. Let G be a set with a binary operation denoted by $*$. Then G is a **group** if the following axioms hold:

(G1) (associativity) For all $g, h, k \in G$, $(g * h) * k = g * (h * k)$.

(G2) (existence of an identity) There exists $e \in G$ such that for all $g \in G$, $e * g = g * e = g$.

(G3) (existence of inverses) For all $g \in G$, there exists an element $h \in G$ such that $g * h = h * g = e$. We call h the **inverse** of g and denote it by g^{-1}.

Definition 6.11. Let G be a group. G is **abelian** if the binary operation is also commutative, that is, if $g * h = h * g$ for all $g, h \in G$.

You have, of course, met many groups already even if you did not call them groups. Let's look at some examples.

Example 6.12. Consider $\mathbb{Z}$ with the binary operation of addition. We know that addition is associative, that 0 is an identity, and that every integer n has an additive inverse $-n$. Thus $\mathbb{Z}$ is a group under addition. Because the operation of addition is also commutative, $\mathbb{Z}$ is an abelian group.

Example 6.13. Consider $\mathbb{Z}_p$ with the binary operation of addition modulo p. We have also seen that this operation is associative, that there is an identity element, and that each element has an additive inverse. Thus $\mathbb{Z}_p$ is a group under addition. If p is prime, $\mathbb{Z}_p \setminus \{\bar{0}\}$ is also a group under multiplication. Because addition and multiplication modulo p are commutative, both groups are abelian.

Exercise 6.26. In the last example, we commented that, if p is prime, $\mathbb{Z}_p \setminus \{\bar{0}\}$ is a group under multiplication. Is $\mathbb{Z}_p$ itself a group under multiplication?

Exercise 6.27. Why isn't $\mathbb{Z}_4 \setminus \{\bar{0}\}$ a group under multiplication?

Our next example may seem a bit contrived, but it turns out that the group described is one of the most important examples.

Example 6.14 (Symmetric Group). Let S_n be the set of all bijections from $\{1, 2, \ldots, n\}$ to $\{1, 2, \ldots, n\}$ with the binary operation of function composition. We know that function composition is associate. Also, the function id defined by $id(k) = k$ for all $k \in \{1, 2, \ldots, n\}$ has the properties of an identity. Finally, because every bijection has a well-defined bijective inverse function, every element of S_n has an inverse that is also in S_n. S_n is called the **symmetric group** on n elements.

Exercise 6.28. Consider S_n as defined in the preceding example.

(a) How many elements are in S_3? List them all. How many elements are in S_n?

(b) Show that S_3 is not abelian.

With these few examples in mind, we prove some very elementary propositions about groups.

Proposition 6.12. *The identity element of a group G is unique.*

We almost always approach uniqueness proofs the same way; we suppose we have two objects with the desired property and show that, in fact, they must be equal.

Proof. Suppose e and e' are two elements of G with the identity property. Then because e' is an identity, $e = e * e'$, but because e is an identity, $e * e' = e'$. Thus $e = e'$, proving that the identity element of G is unique. □

Proposition 6.13. *Let G be a group and let g be any element of G. Then the inverse of g is unique.*

Proof. The proof is left to the reader as the next exercise. □

Exercise 6.29. Prove Proposition 6.13 using the same idea as in the proof of Proposition 6.12.

Proposition 6.14. *Let G be a group and let $g, h, k \in G$. If $g * h = g * k$, then $h = k$.*

Proof. Because G is a group, g has an inverse, g^{-1}. Then

$$h = e * h = (g^{-1} * g) * h = g^{-1} * (g * h) = g^{-1} * (g * k),$$

where we have used the existence of the identity, the existence of inverses, associativity, and finally the hypothesis of the proposition. Using these same properties,

$$g^{-1} * (g * k) = (g^{-1} * g) * k = e * k = k.$$

Thus $h = k$. □

One way to describe a small group is by making a group table. The table lists the elements of the group down the first column and across the first row. The entries then give the result of the binary operation. If we are in the row for element g and the column for element h, the table will display $g * h$ at the intersection of that row and column. We illustrate with a group having three elements.

$*$	e	a	b
e	e	a	b
a	a	b	e
b	b	e	a

Group tables actually have a lot of structure to them. We note first that the first row and first column are completely determined because e is the identity of the group. Second, we note that the identity element must appear in each row and each column because each element of the group must have an inverse. Third, no group element can appear twice in the same row or column. To see why, suppose ℓ appeared twice in the row for element g. It would follow that there exist distinct group elements h and k such that $g * h = \ell = g * k$. By Proposition 6.14, $h = k$. This contradiction leads us to conclude that ℓ can only appear once in the row for g.

These considerations imply that (up to renaming of the elements) there is only one possible table for a group of three elements. There is also only a single group with two elements. A bit of work shows that there are two distinct groups with four elements. You will explore all these assertions in the exercises and problems.

Exercise 6.30. Make the group table for a group with two elements. Explain why only one such group exists.

Exercise 6.31. Consider the following partial table for a group with three elements.

$*$	e	a	b
e	e	a	b
a	a	?	
b	b		

Explain carefully why the question mark must be the element b.

6.5.3 Rings and fields

We now consider some of the possible algebraic structures that arise if we have a set and two binary operations.

Definition 6.12. Let R be a set with two binary operations $+$ and $\cdot$. R is a **ring** if the following axioms hold:

(R1) (associativity of addition) For all $a, b, c \in R$, $(a + b) + c = a + (b + c)$.

(R2) (commutativity of addition) For all $a, b \in R$, $a + b = b + a$.

(R3) (existence of additive identity) There exists $0 \in R$ such that for all $a \in R$, $a + 0 = 0 + a = a$.

(R4) (existence of additive inverses) For all $a \in R$, there exists an element $b \in R$ such that $a + b = b + a = 0$. b is called the **additive inverse** of a and denoted $-a$.

(R5) (associativity of multiplication) For all $a, b, c \in R$, $(a \cdot b) \cdot c = a \cdot (b \cdot c)$.

(R6) (existence of multiplicative identity) There exists $1 \in R$ such that for all $a \in R$, $a \cdot 1 = 1 \cdot a = a$.

(R7) (distributive law) For all $a, b, c \in R$, $a \cdot (b + c) = a \cdot b + a \cdot c$ and $(a + b) \cdot c = a \cdot c + b \cdot c$.

The first four ring axioms can be summarized by saying that, to begin with, R is an abelian group under addition. Note that R is not a group under multiplication because the elements of R do not necessarily have multiplicative inverses in R.

The ring axioms should look familiar; these are almost exactly the properties satisfied by the integers. The only difference is that the multiplication in the integers is also commutative whereas the definition of a ring does not require this property to hold. If the multiplication in R is commutative, we call R a *commutative ring*. The integers furnish the prototypical example of a commutative ring. We give several other examples of rings.

Example 6.15. $\mathbb{Z}_p$ with operations of addition modulo p and multiplication modulo p is another example of a commutative ring.

Example 6.16. Let M_2 be the set of 2×2 matrices with binary operations of matrix addition and matrix multiplication. M_2 is a ring, but it is not commutative.

Exercise 6.32. Consider M_2 as defined in the preceding example.

(a) What are the additive and multiplicative identities of M_2? Show that they have the desired properties.

(b) Let $A \in M_2$. What is its additive inverse?

(c) Show that M_2 is not a commutative ring by finding $A, B \in M_2$ for which $AB \neq BA$.

The final algebraic structure we consider is that of a field. Said simply, a field is a commutative ring in which every non-zero element has a multiplicative inverse. To exclude the trivial ring consisting of only one element, we require that the additive and multiplicative identity be distinct. Although we listed the field axioms in Chapter 1, we include them again for reference.

Definition 6.13. A set F with two binary operations addition $(+)$ and multiplication (denoted by juxtaposition) is a **field** if the following axioms are satisfied:

(F1) (Associativity) For all x, y, z in F, $(x + y) + z = x + (y + z)$ and $(xy)z = x(yz)$.

(F2) (Commutativity) For all x, y in F, $x + y = y + x$ and $xy = yx$.

(F3) (Existence of identities) There exists an element 0 of F such that $x + 0 = 0 + x = x$ for all x in F. Furthermore, there exists an element 1 of F different from 0, such that $x1 = 1x = x$ for all x in F. 0 is the **additive identity** and 1 is the **multiplicative identity**.

(F4) (Existence of inverses) If x is in F, there exists a unique element of F, denoted $-x$, such that $x + (-x) = (-x) + x = 0$. Furthermore, if x is in F and $x \neq 0$, there exists a unique element of F, denoted x^{-1}, such that $xx^{-1} = x^{-1}x = 1$. $-x$ is the **additive inverse** of x and x^{-1} is the **multiplicative inverse** of x.

(F5) (Distributive property) For all x, y, z in F, $x(y+z) = xy+xz$ and $(x+y)z = xz + yz$.

Of course the rational numbers $\mathbb{Q}$ are the prototypical example of a field. The real numbers $\mathbb{R}$ are another familiar example. We discuss two other examples here.

Example 6.17. Define a symbol i by the property that $i^2 = -1$. Let $\mathbb{C} = \{a + ib : a, b \in \mathbb{R}\}$. We define operations on $\mathbb{C}$. Let $z, w \in \mathbb{C}$. Thus there exist

$a, b, c, d \in \mathbb{R}$ such that $z = a + ib$ and $w = c + id$. Define the sum and product of z and w in the expected way.

$$z + w = (a + c) + i(b + d) \tag{6.10}$$
$$zw = (ac - bd) + i(ad + bc). \tag{6.11}$$

In particular, observe that zw is exactly what it has to be if the distributive law is to apply to $(a+ib)(c+id)$. We claim $\mathbb{C}$ is a field, called the *complex numbers.*

Exercise 6.33. Verifying all the field axioms for $\mathbb{C}$ is tedious, so we only ask you to consider a few.

(a) Show that multiplication of complex numbers is commutative.

(b) Show that $1 + i0$ is the multiplicative identity of $\mathbb{C}$.

(c) Show that every non-zero element of $\mathbb{C}$ has a multiplicative inverse. (Suggestion: Given a non-zero element $a + ib$, find a formula for its multiplicative inverse and verify that it works.)

Example 6.18. $\mathbb{Z}_p$ with operations of addition and multiplication modulo p is also a field if p is prime. If p is not prime, $\mathbb{Z}_p$ can have non-zero elements without multiplicative inverses. See Exercise 6.27.

As we indicated in Chapter 1, when a set has the algebraic structure of a field, we may do the same sorts of algebraic manipulations we learned in high school algebra. For example, we can always solve a linear equation.

Proposition 6.15. *Let F be a field and suppose $a, b \in F$. Then $ax + b = 0$ has the unique solution $x = a^{-1}(-b)$.*

Several proofs are possible. One mimics the development in Chapter 1. There, we proved Proposition 1.2, which legitimized the steps we do all the time in algebra. An examination of the proof of that proposition reveals that it relies only on the field axioms. Thus the same proposition holds in any field. A second possible proof simply verifies that the proposed element $x = a^{-1}(-b)$ is a solution to the equation $ax + b = 0$ and then shows that any arbitrary solution must in fact be equal to this one. Both proofs are easy and left as an exercise.

Here is another simple but important proposition about fields.

Proposition 6.16. *Let F be a field and let $a, b \in F$. Then $ab = 0$ if and only if $a = 0$ or $b = 0$.*

Again, we saw this property of the real numbers in Chapter 1, where the proof was left as an exercise. This property is also very frequently used in algebra. For example, suppose you want to solve $x^2 + 5x + 6 = 0$. A common approach is to *factor*, i.e., to write the polynomial as a product of linear factors:

$$x^2 + 5x + 6 = (x + 2)(x + 3).$$

By Proposition 6.16, this product equals 0 if and only if either $x + 2 = 0$ or $x + 3 = 0$. We then solve these linear equations to obtain the solutions $x = -2$ and $x = -3$.

Unfortunately, Proposition 6.16 does not hold in an arbitrary ring.

Example 6.19. Consider $\mathbb{Z}_6$. Then $\bar{2}$ and $\bar{3}$ are non-zero elements, but $\bar{2} \cdot \bar{3} = \bar{0}$.

Definition 6.14. Let R be a ring. A non-zero element a of R is called a **left zero divisor** if there exists a non-zero element b of R such that $ab = 0$. We define **right zero divisors** in a similar manner. If the ring R is commutative, either type of element is simply called a **zero divisor**.

Thus in $\mathbb{Z}_6$, $\bar{2}$ and $\bar{3}$ are zero divisors. The exercises give you an opportunity to explore several other rings to determine whether or not they contain zero divisors.

We have only just scratched the surface in our discussion of algebraic structures. We hope that you see how these structures abstract the properties of familiar number systems. We hope you have also glimpsed the diversity of examples that share some common structure. If you go on to take further courses in abstract algebra, never lose sight of the examples. It can be fun to just see what you can prove from a given set of axioms, but remember that every set of axioms we study is studied because of our desire to understand specific examples. Pure math needn't explore structures in the physical world, but it is always well motivated. It is not just a formal game of defining arbitrary structures and deriving their logical consequences.

6.6 Problems

1. When we learn to add multi-digit numbers in base 10, we learn how to "carry." Do the following arithmetic problems in bases other than 10 without converting the numbers to base 10. Then write a description of how "carrying" works in any base b.

   (a) $121_3 + 202_3$.

   (b) $378_9 + 286_9$.

   (c) $10101_2 + 11111_2$.

2. (a) Let $a, b \in \mathbb{Z}$ and let $d = \gcd(a, b)$. Let $S = \{ma + nb : m, n \in \mathbb{Z}\}$ and let $T = \{kd : k \in \mathbb{Z}\}$. Show that $S = T$.

   (b) Does the equation $12x + 21y = 88$ have any integer solutions?

3. Let n be a natural number. Give at least two distinct proofs that 3 divides $n^3 - n$.

4. Let n be an integer and show that 5 divides $n^5 - 9n^3 + 20n$.

5. The *dihedral group* of the square is the group of all rigid motions of the square. These include rotations and reflections over different axes of symmetry. Describe all the elements of the group. Then make a group table for this group.

6. There are two distinct groups with four elements. Make their group tables.

7. The field axioms require a field to have at least two elements, 0 (the additive identity) and 1 (the multiplicative identity). Let F be a field with only three elements.

   (a) Make the complete addition table for F.

   (b) Make the complete multiplication table for F.

   (c) A quadratic equation in F is an equation of the form $ax^2+bx+c=0$ where $a, b, c \in F$. Determine which quadratic equations in F have solutions in F.

8. Consider M_2, the ring of 2×2 matrices with real entries.

   (a) Is M_2 a field?

   (b) Does M_2 have the property that, if $AB = 0$, either $A = 0$ or $B = 0$?

9. Consider the field $\mathbb{C}$ of complex numbers. Is it possible to define $<$ on $\mathbb{C}$ so that $\mathbb{C}$ is an ordered field? (Hint: Think about trichotomy and the complex number i.)

10. (Challenging) Suppose we have postage stamps with denominations of a cents and b cents, where $a < b$ and a and b are relatively prime.

    (a) Prove that postage of $ab - a - b$ can not be made.

    (b) Let $S = \{jb : 0 \leq j \leq a - 1\}$. Show that distinct elements of S are in distinct congruence classes modulo a.

    (c) Prove that every postage P with $P \geq (a-1)b$ can be made.

    (d) Prove that postage of $P = ab - a - b + 1$ can be made and that, in fact, if $P \geq ab - a - b + 1$, postage P can be made.

Programming Project

Lemma 6.1 shows that $\gcd(a, b) = \gcd(a - b, b)$.

1. Write a computer program that uses this fact to compute $\gcd(a, b)$.

2. Prove that your algorithm returns $\gcd(a, b)$. Model your proof after the proof of the Euclidean algorithm.

Chapter 7

Cardinality

7.1 The Definition

Which is bigger, $\mathbb{N}$ or $\mathbb{Z}$? $\mathbb{Q}$ or $\mathbb{R}$? A good mathematics student will refuse to answer these questions because we have not said what it means for one set to be bigger than another. Certainly $\mathbb{N}$ is a proper subset of $\mathbb{Z}$ and $\mathbb{Q}$ is a proper subset of $\mathbb{R}$. Defining size in terms of containment, however, is probably not the approach we want to take; if $A = \{1, 2, 3, 4\}$ and $B = \{3, 4, 5, 6, 7\}$, we think of B as bigger than A even though A is not a subset of B.

In Chapter 5, for a finite set A, we defined its cardinality $|A|$ to be the number of elements in A. For finite sets A and B, we could agree to say that B is bigger than A if $|B| > |A|$. Thus for the two sets in the preceding paragraph, $|A| = 4$, $|B| = 5$, and so $|B| > |A|$. This definition is fine as far as it goes, but it does not apply to infinite sets. Thus we want to give a better definition of cardinality that both encompasses the definition we gave for finite sets and gives us a way to compare infinite sets.

Definition 7.1. Let A and B be sets. A has the same **cardinality** as B if there exists a bijection $f\colon A \to B$.

By Proposition 4.4, if f is a bijection from A to B, then f^{-1} is a bijection from B to A. Thus "has the same cardinality" is a symmetric relation, and we can just say that A and B have the same cardinality when a bijection exists between them.

Example 7.1. If $A = \mathbb{N}$ and $B = \{n \in \mathbb{Z} : n \leq -1\} = \{-1, -2, -3, \ldots\}$, then we can define $f\colon A \to B$ by $f(n) = -n$. It is very easy to see that f is a bijection, and so A and B have the same cardinality.

Example 7.2. For a less trivial example, consider $\mathbb{Z}$ and $3\mathbb{Z}$. Clearly $3\mathbb{Z}$ is a proper subset of $\mathbb{Z}$. It is easy to see, however, that they have the same cardinality. Indeed, define $f\colon \mathbb{Z} \to 3\mathbb{Z}$ by $f(n) = 3n$. Clearly f is a bijection. This example shows that a set can have the same cardinality as one of its proper subsets.

Example 7.3. Let A be the open interval $(1,2)$ of $\mathbb{R}$ and let B be the open interval $(4,6)$ of $\mathbb{R}$. We claim that A and B have the same cardinality. We will take as our bijection the linear function whose graph is the line passing through the points $(1,4)$ and $(2,6)$. Thus define $f\colon A \to B$ by $f(x) = 2x + 2$. We must show that f is both injective and surjective.

Suppose $f(x_1) = f(x_2)$. Then

$$2x_1 + 2 = 2x_2 + 2 \iff 2x_1 = 2x_2 \iff x_1 = x_2$$

and f is injective. To see that f is surjective, take $y \in B$, so that $4 < y < 6$. Then if $x = \frac{1}{2}y - 1$,

$$f(x) = 2\left(\frac{1}{2}y - 1\right) + 2 = y - 2 + 2 = y.$$

Furthermore,

$$4 < y < 6 \iff 2 < \frac{1}{2}y < 3 \iff 1 < \frac{1}{2}y - 1 < 2.$$

Because every y in B has a pre-image x in A, f is surjective. Thus f is a bijection.

Exercise 7.1. Show that $\mathbb{N}$ and $\mathbb{N} \cup \{0\}$ have the same cardinality by finding a bijection $f\colon \mathbb{N} \to \mathbb{N} \cup \{0\}$.

Exercise 7.2. Show that the open intervals $(0,1)$ and $(1,\infty)$ have the same cardinality.

7.2 Finite Sets Revisited

When we talked about cardinality in Chapter 5, we did not talk about bijections. We simply defined the cardinality of a set as the number of elements in that set. In this section we hope to convince you that these two notions are really the same. We must think more deeply about the process of counting.

What do we do when we count? When we say that there are 7 days in a week, we mean that we have put the set

$$D = \{\text{Sunday, Monday, Tuesday, Wednesday, Thursday, Friday, Saturday}\}$$

in a one-to-one correspondence with the set

$$\{1,2,3,4,5,6,7\},$$

probably pairing Sunday with 1, Monday with 2, and so on. In other words, we have created a bijection between the set to be counted and the known set $\{1,2,3,4,5,6,7\}$. Our experience with counting tells us that the answer is unique, so that, for example, there is no bijection between the set of days in a week and the set $\{1, 2, 3, 4, 5, 6\}$. In other words, we believe that the cardinality of a finite set is *well-defined.*

Should we try to prove this statement? We could get into a lengthy discussion of how to define the natural number system, but this really isn't the direction we want to go in this course. To get a feeling for the issues that might be involved, you might consider the following exercises.

Exercise 7.3. Try to make an argument that there is no bijection between the set D of days of the week and $\{1, 2, 3, 4, 5, 6\}$ without referring to the number of elements in either set. (Such an argument would be circular, for the question at hand is whether the number of elements in a set is well-defined.)

Exercise 7.4. Let $S = \{a, b\}$, $T = \{c, d\}$, and $U = \{e, f, g\}$.

1. Show that S and T have the same cardinality by finding a bijection between them.

2. Show that S and U do not have the same cardinality by showing that there is no bijection between them. Your argument may not refer to the number of elements in either of the sets.

As we have done from the beginning of this book, we will continue to assume that we know the natural numbers and that we know how to count finite sets. In other words, we continue to assume that cardinality is well-defined for finite sets. For completeness, we give this definition. In it, we use $[n]$ to denote the set $\{j : 1 \leq j \leq n\}$.

Definition 7.2. Let A be a non-empty set. A is **finite** if there is a bijection between A and $[n]$ for some natural number n. We say that the cardinality of A is n and write $|A| = n$.

Of course we define the cardinality of the empty set to be 0.

We state some simple but important properties of finite sets. Perhaps the most important reason to state these properties is that the infinite sets we study later in this chapter need not satisfy these properties.

Proposition 7.1. *Let A be a non-empty set and suppose $|A| = n$. Let $a \in A$ and define a new set $B = A \setminus \{a\}$. Then $|B| = n - 1$.*

Proof. Because $|A| = n$, there exists a bijection $f : A \to [n]$. Suppose $f(a) = J$. Now consider B. If $n = 1$, A is the singleton set $\{a\}$, $B = \emptyset$, and $|B| = 0$, as claimed. We therefore suppose for the remainder of the proof that $n > 1$ so that B is non-empty.

The restriction of f to B, denoted $f|_B$, is a bijection from B to $[n] \setminus \{J\}$. We define $g : [n] \setminus \{J\} \to [n-1]$ by setting

$$g(j) = \begin{cases} j & 1 \leq j < J \\ j - 1 & J < j \leq n. \end{cases}$$

Clearly g is a bijection. Then $g \circ f|_B : B \to [n-1]$ is a bijection because it is a composition of bijections. We conclude that $|B| = n - 1$. □

Proposition 7.2. *Let A and B be finite sets with $|A| = |B|$.*

(a) If $f: A \to B$ is injective, then f is in fact surjective.

(b) If $f: A \to B$ is surjective, then f is in fact injective.

Proof. We prove part (a) and leave the proof of part (b) as an exercise.

Suppose $|A| = |B| = n$. Because f is injective, f gives a bijection between A and its image $f(A)$. Thus A and $f(A)$ have the same cardinality, that is, $|f(A)| = |A| = n$. Because $f(A) \subseteq B$ and both $f(A)$ and B have n elements, $f(A) = B$. In other words, f is surjective. □

Exercise 7.5. Prove part (b) of Proposition 7.2

Proposition 7.3. *Let A be a non-empty finite subset of an ordered set S. Then A has a largest and a smallest element.*

Proof. We prove that A has a largest element; the proof that A has a smallest element is nearly identical.

Our proof is by induction on n, the cardinality of A. If $|A| = 1$, then $A = \{a\}$ and a is trivially the largest element of A. Suppose now that the result holds for n, i.e., that any set with cardinality n has a largest element. Consider now a set A with $|A| = n + 1$. Because A is non-empty, we may select an element a of A. Form a new set $B = A \setminus \{a\}$. By Proposition 7.1, $|B| = n$, and so by the inductive hypothesis, B has a largest element, which we denote M. Now consider a and M. Because every pair of elements is comparable in an ordered set, either $a < M$ and M is the largest element of A or $M < a$ and a is the largest element of A. In either case, A has a largest element, so the result holds for sets with cardinality $n + 1$. By the principle of mathematical induction, the result holds for all n. □

7.3 Countably Infinite Sets

Definition 7.3. A set A is called **infinite** if it is non-empty and for all $n \in \mathbb{N}$ there is no bijection between A and $[n]$.

The most obvious example of an infinite set is $\mathbb{N}$ itself. Let's prove that, for all $n \in \mathbb{N}$, there is no bijection between $\mathbb{N}$ and $[n]$. We give a proof by contradiction. Suppose, then, that for some $n \in \mathbb{N}$ there is a bijection $f: [n] \to \mathbb{N}$. Consider $f([n])$. This is a finite set with n elements because it is in bijective correspondence with $[n]$. Because $f([n])$ is a subset of the ordered set $\mathbb{N}$, by Proposition 7.3 it has a largest element N. But then $N + 1$ is an element of $\mathbb{N}$ that is larger. Hence $N + 1$ is not in the image of f, and f is not surjective. We have reached a contradiction. We conclude that no such bijection exists.

Not only is $\mathbb{N}$ a simple example of an infinite set, but it is also used to define an entire class of infinite sets.

Definition 7.4. Let A be a non-empty set. A is **countably infinite** if A has the same cardinality as $\mathbb{N}$, that is, if there exists a bijection $f: A \to \mathbb{N}$.

We will sometimes refer to a set that is either finite or countably infinite as a *countable* set. For our first example, we return to the question with which we opened the chapter.

Example 7.4. We show that $\mathbb{Z}$ is countably infinite. Our definition requires us to construct a bijection from $\mathbb{Z}$ to $\mathbb{N}$. Define such a function by the rule

$$f(n) = \begin{cases} 2n+1 & n \geq 0 \\ -2n & n < 0. \end{cases}$$

This function maps the non-negative integers onto the odd natural numbers and maps the negative integers onto the even natural numbers. It is not hard to see that f is a bijection. We omit the details.

In other words, Example 7.4 shows that $\mathbb{N}$ and $\mathbb{Z}$ have the same cardinality. In this sense, these sets have the same size. This fact may strike you as counterintuitive; it seems as if there are "twice as many" integers as natural numbers. This example shows again why definitions are so important. Our definition says two sets have the same cardinality if there is a bijection between them. These two sets have a bijection between them, so they have the same cardinality whether or not this fact seems intuitive. You may find many results about infinite sets counterintuitive, and you may catch yourself writing arguments that are wrong because you assume something is true about infinite sets because it is true for the more familiar finite sets. We encourage you to stick to your definitions. They will tell you how to proceed. Over time, you will develop a new intuition for infinite sets.

When a set A is countably infinite, it means we can essentially list the elements. Indeed, because there is a bijection between A and $\mathbb{N}$, for every $n \in \mathbb{N}$ there is an element of A paired with it. We call this element a_n. We then say that $a_1, a_2, a_3, \ldots$ is an *enumeration* of A. In Example 7.4 we obtained the following enumeration of the integers:

$$a_1 = 0, a_2 = -1, a_3 = 1, a_4 = -2, a_5 = 2, \ldots.$$

Exercise 7.6. Recall that a sequence is a function from $\mathbb{N}$ or $\mathbb{N} \cup \{0\}$ into some set A. Is every sequence with values in A an enumeration of A? Give proofs or counterexamples.

We now prove a basic proposition about unions of infinite sets. Later in the problems you will consider several possible generalizations of this result.

Proposition 7.4. *Let A and B be countably infinite and disjoint. Then $A \cup B$ is countably infinite.*

Proof. Because A and B are countably infinite, we may obtain an enumeration of each. Thus let $a_1, a_2, a_3, \ldots$ be an enumeration of A and let $b_1, b_2, b_3, \ldots$ be an enumeration of B. Define $f \colon A \cup B \to \mathbb{N}$ by $f(a_n) = 2n - 1$ and $f(b_n) = 2n$. In other words, the elements of A are mapped onto the odd natural numbers and the elements of B are mapped onto the even natural numbers. We omit the details of the argument that f is a bijection. $\square$

With this proposition established, we are ready to consider a set that seems *much* bigger than the natural numbers, namely the rational numbers $\mathbb{Q}$. Could this set possibly be countable? It turns out it is. We will give two arguments in this section and suggest a third in the problems at the end of the chapter. The first appeals directly to the definition.

Example 7.5. Let $\mathbb{Q}_+$ denote the set of positive rational numbers. We will prove that $\mathbb{Q}_+$ is countably infinite. The result for $\mathbb{Q}$ itself then follows by writing $\mathbb{Q} = \mathbb{Q}_+ \cup \mathbb{Q}_- \cup \{0\}$ and applying Proposition 7.4.

We begin by writing all ratios of integers in an infinite array in which all entries in a row have the same denominator and all entries in a column have the same numerator:

$$\begin{array}{ccccc} \frac{1}{1} & \frac{2}{1} & \frac{3}{1} & \frac{4}{1} & \cdots \\ \frac{1}{2} & \frac{2}{2} & \frac{3}{2} & \frac{4}{2} & \cdots \\ \frac{1}{3} & \frac{2}{3} & \frac{3}{3} & \frac{4}{3} & \cdots \\ \frac{1}{4} & \frac{2}{4} & \frac{3}{4} & \frac{4}{4} & \cdots \\ \vdots & \vdots & \vdots & \vdots & \ddots \end{array}$$

We construct our bijection with $\mathbb{N}$ as follows: Begin in the upper left-hand corner of the matrix with the entry a_{11}, then proceed to a_{12} and a_{21}, then a_{13}, a_{22}, a_{31}, etc. Perhaps this figure will help:

$$\begin{array}{ccccc} \swarrow & \swarrow & \swarrow & \swarrow & \cdots \\ \swarrow & \swarrow & \swarrow & \swarrow & \cdots \\ \swarrow & \swarrow & \swarrow & \swarrow & \cdots \\ \swarrow & \swarrow & \swarrow & \swarrow & \cdots \\ \vdots & \vdots & \vdots & \vdots & \cdots. \end{array} \tag{7.1}$$

For each entry, if the fraction is not equivalent to a fraction we have already encountered, we assign it to the next natural number. If it is equivalent to one we have already encountered, we cross it out and make no assignment. In this manner we obtain an enumeration of the positive rationals. This list begins

$$1, 2, \frac{1}{2}, 3, \frac{1}{3}, 4, \frac{3}{2}, \ldots.$$

Exercise 7.7. To make sure you understand the last example, find the next seven fractions in this enumeration of the positive rationals.

Exercise 7.8. Use a modification of the argument in Example 7.5 to prove that $\mathbb{N} \times \mathbb{N}$ is countably infinite.

Sometimes finding an explicit bijection is hard, but finding an injection is easy. The next proposition tells us that doing so is enough.

Proposition 7.5. *Let A be an infinite set and suppose $f\colon A \to \mathbb{N}$ is injective. Then in fact there is a bijection from A to $\mathbb{N}$.*

Proof. We observe first that because f is injective, it establishes a bijection between A and $f(A)$. Let $B_1 = f(A)$. Because B_1 is a non-empty subset of $\mathbb{N}$, by the well-ordering principle, B_1 has a smallest element. Call it b_1, and set $B_2 = B_1 \setminus \{b_1\}$. Note that B_2 is still an infinite set. Proceed inductively, so that having selected $b_1, \ldots, b_{n-1}$, at step n we take b_n to be the smallest element of B_n and then set $B_{n+1} = B_n \setminus \{b_n\}$. In this manner we obtain an enumeration of $f(A)$. In other words, each element of $f(A)$ is equal to b_n for some n, and $g(b_n) = n$ is a bijection between $f(A)$ and $\mathbb{N}$. Then $g \circ f\colon A \to \mathbb{N}$ is a composition of bijections, hence is a bijection. □

We will use this proposition to give a different, shorter argument showing that the positive rationals are countably infinite.

Example 7.6. The positive rational numbers are in a one-to-one correspondence with the set A of ratios $\frac{m}{n}$ of natural numbers for which m and n have no common factors. Now define a map $g\colon A \to \mathbb{N}$ by

$$g\left(\frac{m}{n}\right) = 2^m 3^n.$$

The Fundamental Theorem of Arithmetic implies that if $2^m 3^n = 2^M 3^N$, then $m = M$ and $n = N$. Thus g is injective. Therefore by Proposition 7.5, there is a bijection between A and $\mathbb{N}$. Because $\mathbb{Q}_+$ and A are in bijective correspondence, there is a bijection between $\mathbb{Q}_+$ and $\mathbb{N}$. Therefore $\mathbb{Q}_+$ is countably infinite.

A simple modification of this argument establishes the following proposition concerning Cartesian products.

Proposition 7.6. *Let A and B be countably infinite sets. Then $A \times B$ is countably infinite.*

Proof. Because A and B are countably infinite, we may obtain an enumeration of each. Thus let $a_1, a_2, a_3, \ldots$ be an enumeration of A and let $b_1, b_2, b_3, \ldots$ be an enumeration of B. Every element of $A \times B$ can be written (a_m, b_n) for some $m, n \in \mathbb{N}$. Define $f\colon A \times B \to \mathbb{N}$ by the rule $f(a_m, b_n) = 2^m 3^n$. As we pointed out above, this map is injective, and so by Proposition 7.5 $A \times B$ is countably infinite. □

7.4 Uncountable Sets

Are all infinite sets countably infinite? The previous section certainly contained some surprises–sets that we think of as much bigger than $\mathbb{N}$, like $\mathbb{Q}$, that turn out to be countably infinite. In this section, we give an example of an *uncountable* set.

Example 7.7. Let S be the set of all infinite sequences of zeros and ones. We will prove that S is not countably infinite. We will give a proof by contradiction.

Suppose S is countably infinite. Then there is a one-to-one correspondence between S and $\mathbb{N}$, hence an enumeration $s_1, s_2, \ldots, s_n, \ldots$ of S. Now each s_n is a sequence of zeros and ones, that is, $s_n = \{a_{nm}\}$, where each a_{nm} is either 0 or 1. We arrange our infinite collection of infinite sequences in an array so that the first row is the sequence s_1, the second row is the sequence s_2, and so on.

$$\begin{array}{cccc} a_{11} & a_{12} & a_{13} & \cdots \\ a_{21} & a_{22} & a_{23} & \cdots \\ a_{31} & a_{32} & a_{33} & \cdots \\ \vdots & \vdots & \vdots & \ddots \end{array} \tag{7.2}$$

We now use this array to construct a new sequence $b = \{b_n\}$ that is not on the list. To define b_n, consider a_{nn}. If $a_{nn} = 0$, set $b_n = 1$. If $a_{nn} = 1$, set $b_n = 0$. This construction guarantees that sequence b differs from sequence s_n in at least one term. Now b is a sequence of zeros and ones, so it must be in S, but it is not equal to any of the s_n. We have reached a contradiction, for we supposed $n \mapsto s_n$ was a one-to-one correspondence between S and $\mathbb{N}$. Therefore no such correspondence exists and S is not countably infinite.

Definition 7.5. Let A be an infinite set and suppose there does not exist a bijection between A and $\mathbb{N}$. Then A is **uncountable**.

Exercise 7.9. Let A and B be infinite sets with $A \subseteq B$. Suppose that A is countable and B is uncountable. Prove that $B \setminus A$ is uncountable.

Let's think more about Example 7.7. The proof is very clever, but it might not yet be clear to you that the result is also very important; it might seem on first reading as if the set of infinite strings of zeros and ones isn't something we run into all that often. But it turns out that we are most of the way to showing that the interval $[0, 1]$ of real numbers is uncountable. We can only give a sketch of the rest of the argument because it relies on a few facts we have not yet proved. The first fact is that each element of $[0, 1]$ has a binary representation $0.b_1b_2b_3\ldots$, where each b_n is a zero or a one. Such a binary representation is really shorthand for an infinite series $\sum_{n=1}^{\infty} b_n 2^{-n}$. If each element of $[0, 1]$ had a unique binary representation, it would follow immediately from Example 7.7 that $[0, 1]$ is uncountable. Such is not the case. For example, one can show that $.1\bar{0} = .0\bar{1}$. Both are binary representations for $\frac{1}{2}$. This phenomenon is identical to what we see with decimal representations, where $.5\bar{0} = .4\bar{9}$. It turns out that the only numbers with more than one representation are rational numbers, and there are only countably many of those. Thus the fact that the set of infinite strings of zeros and ones is uncountable implies that the set $[0, 1]$ is uncountable.

We have now seen two different sizes of infinite sets; we have seen sets that are the size of $\mathbb{N}$ and sets that are the size of $\mathbb{R}$. One may ask if there are other sizes for infinite sets. The answer is yes, but we will not pursue this topic. You might find it interesting to know that it is still not known whether there is an "intermediate" kind of infinity between $\mathbb{N}$ and $\mathbb{R}$. Phrased precisely, the question would be this: Suppose $\mathbb{N} \subset A \subset \mathbb{R}$. If A does not have the same cardinality as $\mathbb{N}$, must it have the same cardinality as $\mathbb{R}$? The *continuum hypothesis* states that the answer is yes, but it is not known whether it is true.

7.5 Problems

1. Hilbert's hotel has infinitely many rooms, numbered $1, 2, 3, \ldots$ using the natural numbers. On a particular stormy night, each room has precisely one occupant.

   (a) A weary traveller comes to the desk and asks for a room. Neither the traveller nor any guest is willing to share a room, and yet the clerk says he can accommodate the traveller. How can he do so?

   (b) Hilbert's bus has seats numbered $1, 2, 3, \ldots$ with the natural numbers. Every seat is occupied. The driver goes into Hilbert's hotel and says that each passenger would like a room. Every room in the hotel still has an occupant, and no one is willing to share. Can the clerk accommodate the passengers? (Hint: Have you seen these problems before?)

2. Is the result in Proposition 7.4 still true if A and B are not disjoint? If not, why not? If so, how must you modify the argument?

3. Let $\{A_j : j \in \mathbb{N}\}$ be a countable collection of sets, each of which is countably infinite. Suppose for simplicity that the sets are pairwise disjoint (i.e., $A_i \cap A_j = \emptyset$ for all natural numbers i, j with $i \neq j$). Show that $\bigcup_{j=1}^{\infty} A_j$ is countable. Then use the result to give another proof that the set $\mathbb{Q}_+$ of positive rational numbers is countably infinite.

4. Let $\{A_j : 1 \leq j \leq n\}$ be a finite collection of countably infinite sets. Is $A \times A_2 \times \ldots A_n$ countably infinite? Prove it or find a counterexample.

5. Let $\{A_j : j \in \mathbb{N}\}$ be a countably infinite collection of sets. The Cartesian product of this countable collection of sets consists of all infinite sequences $\{a_j\}$ in which $a_j \in A_j$. Under what, if any, conditions on the sets A_j is this Cartesian product countable? Prove your assertions.

6. Let S be the set of all functions from $\mathbb{N}$ to $\{0, 1, 2\}$. Is S countable or uncountable?

7. Determine whether each statement is true or false. If it is true, prove it. If it is false, give a counterexample.

   (a) If A and B are finite sets, $A \cap B$ is finite.

   (b) If A and B are countably infinite sets and $A \cap B$ is not empty, $A \cap B$ is countably infinite.

(c) If A is finite and $A \subseteq B$, then B is finite.

(d) If A is infinite and $A \subseteq B$, then B is infinite.

(e) If A is countably infinite and $A \subseteq B$, then B is countably infinite.

8. Each of the following statements is true for finite sets. Determine whether or not each is true for infinite sets. Give proofs or counterexamples.

   (a) If A is a subset of an ordered set S, then A has a largest element.

   (b) Let A be a non-empty set and let $a \in A$. Let $B = A \setminus \{a\}$. Then A and B do not have the same cardinality.

9. Determine (with proof) whether each of the following sets is finite, countably infinite, or uncountable. If the set is finite, determine the number of elements it has.

   (a) The set of all 2×2 matrices with integer entries.

   (b) The set of all functions from $[n] \to [m]$.

   (c) The set of all functions from $\{0, 1\}$ to $\mathbb{N}$.

   (d) The set of all functions from $\mathbb{N}$ to $\{0, 1\}$.

   (e) The set of all bijections from $[n]$ to $[n]$.

Part II

Foundations of Analysis

Analysis is the branch of mathematics that grows out of calculus and the study of real-valued functions of a real variable. It seeks to make rigorous all the notions of calculus (limit, continuous functions, integrals) and to generalize them to other settings. Thus modern analysis considers functions from $\mathbb{R}^n$ to $\mathbb{R}^m$, complex-valued functions of one or more complex variable, generalizations of the notions of derivative and integral, and vector spaces of functions and the mappings between them.

The aims of this part of the book are to explore the most fundamental notions in analysis and to gain some familiarity with the kinds of arguments used in the field. We thus focus on two main topics: limits of sequences and continuous functions.

Chapter 8

Sequences of Real Numbers

In this chapter, we assume the existence of the ordered field $\mathbb{R}$ of real numbers, although we do not yet discuss or use the completeness of the real numbers. In the next chapter, we will discuss one way to construct the real number system from the rational number system. The reader who wishes to wait to use the real number system until it has been rigorously defined can essentially replace every occurrence of the word "real" in this chapter with the word "rational." We have already carefully considered the ordered field properties for $\mathbb{Q}$, and all the properties of the real number system used in this chapter are also properties of the rational number system.

8.1 The Limit of a Sequence

You probably first studied sequences in second-semester calculus as a prelude to learning about series of numbers, power series, and Taylor's Theorem. At that time, you probably did not work with a formal definition of limit but rather were content with an intuitive understanding. The intuitive definition of limit says that the sequence $a = \{a_n\}$ has limit L if the terms of a are as close as we please to L for n sufficiently large. If a has a limit, we say it is **convergent**. Otherwise we say a is **divergent**. We want to turn this idea into a precise definition that makes explicit what condition must be satisfied by a convergent sequence. Before we give this precise definition, we pause to look at a few examples and to develop some tools for exploring sequences.

8.1.1 Numerical and graphical exploration

In this section, we explore two sequences, a and b, given by n-th term formulas

$$a_n = \frac{(-1)^n n}{2n+1}, \quad n \geq 0$$
$$b_n = \frac{3n}{2n+1}, \quad n \geq 0.$$

One way to explore a sequence is to generate some of its term. At this point, it is a triviality for us to write a few lines of code for this purpose. For sequence a, we might write:

```
N = 20 #the number of terms we'll generate

a = [] #start with an empty list

for n in range(0,N):
    a.append(((-1.0)**n)*n/(2.0*n+1.0))
print(a)
```

Run this code for yourself to generate all 20 terms. The last 4 (rounded) are

$$a[16] = 0.4848485, \quad a[17] = -0.4857143,$$
$$a[18] = 0.4864865, \quad a[19] = -0.4871795.$$

The terms with even index appear to be approaching 0.5 and those with odd index appear to be approaching -0.5. Because there does not appear to be a single number that the terms approach, we suspect that this sequence does not have a limit.

We can look at the same information graphically. To plot this sequence, we simply add two lines to our code:

```
plot(a,".") #the "." tells the program to plot discrete points
show()
```

Figure 8.1 shows the graph.

On the other hand, if we generate 20 terms of the sequence b, the last four (rounded) are

$$b[16] = 1.4545455, \quad b[17] = 1.4571429,$$
$$b[18] = 1.4594595, \quad b[19] = 1.4615385.$$

Figure 8.2 shows the graph of the first 20 terms of b. For sequence b, the numerical and graphical evidence suggests that the terms will be as close to 1.5 as we like for n sufficiently large.

Exercise 8.1. Write your own code to generate the first 20 terms of the sequence b and the graph.

We move towards a more quantitative approach. Let's figure out how large n must be in order for $b[n]$ to be within $\frac{1}{100}$ of $\frac{3}{2}$. That is, we want

$$\left| \frac{3n}{2n+1} - \frac{3}{2} \right| < \frac{1}{100}.$$

Figure 8.1: First 20 terms of the sequence a.

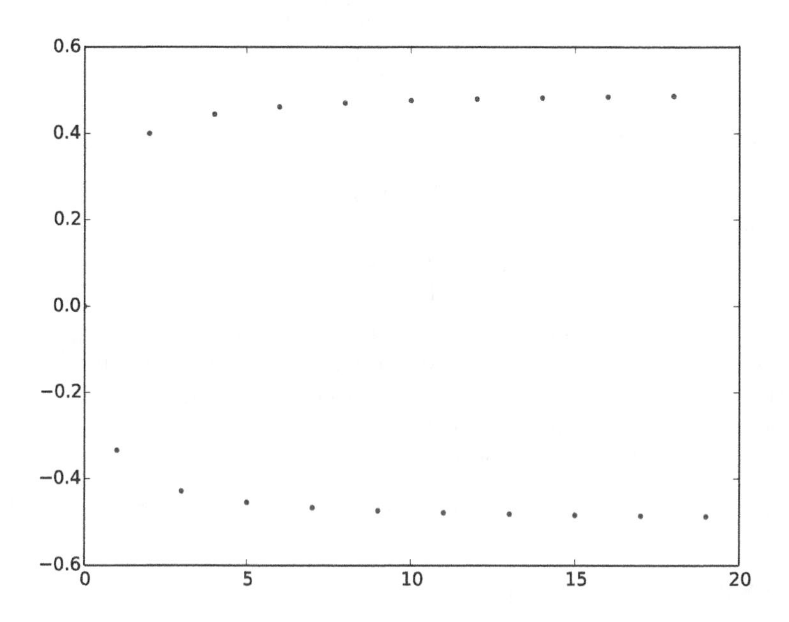

Figure 8.2: First 20 terms of the sequence b.

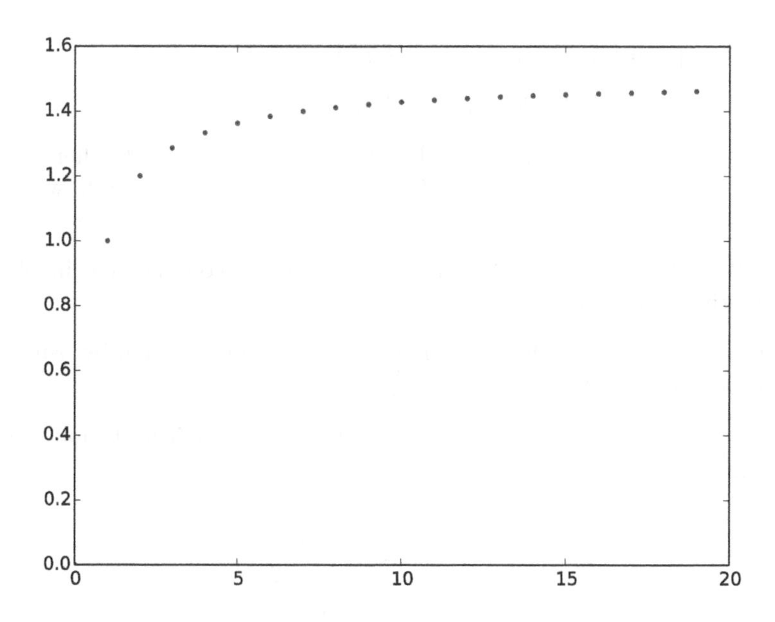

Because

$$\frac{3n}{2n+1} - \frac{3}{2} = \frac{6n - 3(2n+1)}{2(2n+1)} = \frac{-3}{4n+2},$$

we must find n so that

$$\frac{3}{4n+2} < \frac{1}{100}.$$

Thus we require

$$n > \frac{288}{4} = 72.$$

To summarize, we have shown that the terms of b will be within $\frac{1}{100}$ of the (suspected) limit $\frac{3}{2}$ if n is at least 73. We could do this calculation for "errors" ε other than $\frac{1}{100}$. The next exercise asks you to do the calculation for $\varepsilon = \frac{1}{500}$. If for *any* error $\varepsilon > 0$ we can find an N so that, for all $n \geq N$, $\left|b[n] - \frac{3}{2}\right| < \varepsilon$, we will have shown that the sequence indeed has limit $\frac{3}{2}$.

Exercise 8.2. For sequence b above, take $\varepsilon = \frac{1}{500}$ and find an N such that $\left|b[n] - \frac{3}{2}\right| < \varepsilon$ if $n \geq N$.

Exercise 8.3. Consider the sequence with $a_n = 2^{-n}$.

(a) This sequence is decreasing. Write the one-line proof of this fact.

(b) It is not hard to guess that this sequence has limit 0. For each of $\varepsilon = \frac{1}{5}$, $\frac{1}{50}$, $\frac{1}{500}$, find a value of N such that, for all $n \geq N$, $|2^{-n} - 0| < \varepsilon$.

8.1.2 The precise definition of a limit

We have thus motivated the definition of limit we will use.

Definition 8.1. The sequence a has **limit** L if, for every $\varepsilon > 0$, there exists $N \in \mathbb{N}$ such that, for all $n \geq N$, $|a_n - L| < \varepsilon$. When $\{a_n\}$ has limit L, we write $\lim_{n\to\infty} a_n = L$.

Exercise 8.4. The statement defining a limit is quite complicated, involving nested quantifiers. Find its negation.

Example 8.1. Let's finish the example from the previous section by using the definition to prove that $\lim_{n\to\infty} \frac{3n}{2n+1} = \frac{3}{2}$.

Proof. Let $\varepsilon > 0$ be given. We must show that there exists N such that, for all $n \geq N$, $\left|\frac{3n}{2n+1} - \frac{3}{2}\right| < \varepsilon$. Observe,

$$\left|\frac{3n}{2n+1} - \frac{3}{2}\right| = \frac{3}{4n+2} < \frac{3}{4n}.$$

This last expression will be less than ε if $n > \frac{3}{4\varepsilon}$. We may thus take N to be the smallest natural number greater than $\frac{3}{4\varepsilon}$. □

We make an important remark about this proof: The definition of a limit does not require us to find the smallest N that works for a given ε; it only requires us to show that *some* N *exists.* Thus, although we may get a smaller N by solving the inequality

$$\frac{3}{4n+2} < \varepsilon$$

exactly (as we did above in the special case of $\varepsilon = \frac{1}{100}$), we chose instead to find the smallest n for which *the larger expression* $\frac{3}{4n}$ is less than ε. This idea is used very often in analysis proofs and is thus worth understanding thoroughly.

Exercise 8.5. Consider $a_n = \frac{2n-2}{n^2+4}$. Note that $a_n \geq 0$ for all $n \geq 1$.

(a) Find a simple expression b_n such that $0 \leq a_n \leq b_n$ for all $n \geq 1$.

(b) Now take $\varepsilon > 0$. Find an N such that, for all $n \geq N$, $0 \leq a_n < \varepsilon$. The idea, as discussed above, is to find an N so that, for all $n \geq N$, the larger expression b_n is less than ε.

These proofs take some getting used to, so we consider another example.

Example 8.2. We claim $\lim_{n\to\infty} \frac{n}{\sqrt{n^2+1}} = 1$.

Proof. Let $\varepsilon > 0$ be given. We must show that there exists N such that, for all $n \geq N$, $\left|\frac{n}{\sqrt{n^2+1}} - 1\right| < \varepsilon$. Suppose $n \geq 1$. Observe,

$$\begin{aligned}
\left|\frac{n}{\sqrt{n^2+1}} - 1\right| &= \left|\frac{n-\sqrt{n^2+1}}{\sqrt{n^2+1}}\right| \\
&< \left|\frac{n-\sqrt{n^2+1}}{\sqrt{n^2}} \cdot \frac{n+\sqrt{n^2+1}}{n+\sqrt{n^2+1}}\right| \\
&= \left|\frac{-1}{n(n+\sqrt{n^2+1})}\right| \\
&< \frac{1}{n^2},
\end{aligned}$$

where the last inequality comes from replacing $\sqrt{n^2+1}$ in the denominator with the smaller quantity 0. The last expression is less than ε if $n^2 > \frac{1}{\varepsilon}$, or, equivalently, if $n > \frac{1}{\sqrt{\varepsilon}}$. We may thus take N to be any natural number greater than $\frac{1}{\sqrt{\varepsilon}}$. □

Again, we summarize the logic of the argument: We find an N such that, for all $n \geq N$, the larger expression $\frac{1}{n^2}$ is less than ε. Thus the smaller expression $\left|\frac{n}{\sqrt{n^2+1}} - 1\right|$ is also less than ε.

Exercise 8.6. Use the definition of a limit to prove that $\lim_{n\to\infty} \frac{2-n}{n+1} = -1$.

Your intuition tells you that, when a sequence has a limit, that limit is unique. We end this section by proving this statement.

Proposition 8.1. *The limit of a convergent sequence of real numbers is unique.*

Proof. We take our usual approach to proving uniqueness results. Suppose a is a convergent sequence with limits L and M. We show that $L = M$. Suppose not. Then $|L - M| > 0$. Consider $\varepsilon = |L - M|$. For this positive ε, there exist $N_1, N_2 \in \mathbb{N}$ such that for all $n \geq N_1$, $|a_n - L| < \frac{\varepsilon}{2}$, and for all $n \geq N_2$, $|a_n - M| < \frac{\varepsilon}{2}$. Then for all $n \geq N = \max\{N_1, N_2\}$,

$$\begin{aligned} |L - M| &= |(L - a_n) + (a_n - M)| \\ &\leq |L - a_n| + |a_n M| \quad \text{(triangle inequality)} \\ &< \frac{\varepsilon}{2} + \frac{\varepsilon}{2} \\ &= \varepsilon \\ &= |L - M|. \end{aligned}$$

Because we have reached a contradiction, we conclude that $L = M$. □

8.2 Properties of Limits

In a calculus course, you probably learned to do limit calculations algebraically.

Example 8.3. Let's find $\lim_{n\to\infty} \frac{n-3n^2}{2n^2+2}$ using the methods from calculus. The idea is to multiply by a convenient form of 1:

$$\begin{aligned} \lim_{n\to\infty} \frac{n - 3n^2}{2n^2 + 2} &= \lim_{n\to\infty} \frac{n - 3n^2}{2n^2 + 2} \cdot \frac{\frac{1}{n^2}}{\frac{1}{n^2}} \\ &= \lim_{n\to\infty} \frac{\frac{1}{n} - 3}{2 + \frac{2}{n^2}} \\ &= \frac{0 - 3}{2 + 0} = -\frac{3}{2}. \end{aligned}$$

This calculation uses several propositions that we have not yet proved. Most importantly, it uses the fact that *the limit of the quotient is the quotient of the limits* if both limits exist and the limit of the denominator is not zero. The goal of this section is to prove such results so that we can do limit calculations more easily.

Exercise 8.7. Contrast the two approaches by using the definition of a limit to prove that $\lim_{n\to\infty} \frac{n-3n^2}{2n^2+2} = -\frac{3}{2}$.

Exercise 8.8. Example 8.3 also used the fact that, if A is any constant, $\lim_{n\to\infty} \frac{A}{n} = 0$. Prove this fact using the definition of a limit from the previous section. Note that A could be zero. See if you can write a proof that does not treat the case $A = 0$ separately.

Our first proposition will be used in the proofs of some of the limit laws. It is also useful in its own right, for it gives a necessary condition for convergence of a sequence.

Proposition 8.2. *Recall that a sequence a is bounded if there exists M such that $|a_n| \leq M$ for all n. If a is convergent, then a is bounded.*

Proof. The idea of the argument is simple. Because the sequence converges, its tail remains close to L and is hence bounded. There are only finitely many terms not included in this tail, and so these are bounded as well. Taking a maximum gives a bound for the entire sequence.

Here are the details. Let L be the limit of the sequence. Then for $\varepsilon = 1$, there exists N such that, for all $n \geq N$, $|a_n - L| < 1$. Thus by the reverse triangle inequality (Proposition 1.5), for all $n \geq N$, $|a_n| \leq |L| + 1$. Let $M = \max\{|a_0|, \ldots, |a_{N-1}|, |L| + 1\}$. For all n, $|a_n| \leq M$. $\square$

Exercise 8.9. Draw a figure illustrating the idea of this proof.

Exercise 8.10. State the contrapositive of Proposition 8.2. Use it to prove that the sequence with n-th term $a_n = n^2$ is divergent.

Proposition 8.3. *Let a and b be convergent sequences with $\lim_{n\to\infty} a_n = L$ and $\lim_{n\to\infty} b_n = K$. Let A be a real constant.*

(a) The sequence $\{Aa_n\}$ is convergent with $\lim_{n\to\infty} Aa_n = AL$.

(b) The sequence $\{a_n + b_n\}$ is convergent with $\lim_{n\to\infty}(a_n + b_n) = L + K$.

(c) The sequence $\{a_n b_n\}$ is convergent with $\lim_{n\to\infty} a_n b_n = LK$.

(d) If $K \neq 0$, the sequence $\left\{\frac{a_n}{b_n}\right\}$ is convergent with $\lim_{n\to\infty} \frac{a_n}{b_n} = \frac{L}{K}$.

Proof. We leave parts (a) and (d) as exercises and prove parts (b) and (c).
Proof of (b). Let $\varepsilon > 0$ be given. We must show that there exists an N such that, for all $n \geq N$, $|(a_n + b_n) - (L + K)| < \varepsilon$. Because $\lim_{n\to\infty} a_n = L$, for the positive number $\frac{\varepsilon}{2}$, there exists N_1 such that, for all $n \geq N_1$, $|a_n - L| < \frac{\varepsilon}{2}$. Similarly, there exists N_2 such that, for all $n \geq N_2$, $|b_n - K| < \frac{\varepsilon}{2}$. Thus for all $n \geq N = \max\{N_1, N_2\}$, by the triangle inequality

$$\begin{aligned} |(a_n + b_n) - (L + K)| &= |(a_n - L) + (b_n - K)| \\ &\leq |a_n - L| + |b_n - K| \\ &< \frac{\varepsilon}{2} + \frac{\varepsilon}{2} = \varepsilon. \end{aligned}$$

Thus $\{a_n + b_n\}$ converges, with limit $L + K$.
Proof of (c). Let $\varepsilon > 0$ be given. We must show that there exists N such that for all $n \geq N$, $|a_n b_n - LK| < \varepsilon$. We don't know how to relate $a_n b_n$ directly to LK, but we do know something about the relationship between a_n and L.

This fact motivates our next step, i.e., adding and subtracting Lb_n inside the absolute value sign. Observe,

$$\begin{aligned} |a_n b_n - LK| &= |a_n b_n + (-Lb_n + Lb_n) - LK| \\ &= |(a_n - L)b_n + L(b_n - K)| \\ &\leq |a_n - L||b_n| + |L||b_n - K|. \end{aligned}$$

As in the proof of (b), we'd like to use the convergence of a to L to argue that the first term is less than $\frac{\varepsilon}{2}$ for sufficiently large n, and similarly for the second term. Indeed, note first that because b is convergent, by Proposition 8.2, it is bounded. Thus there exists $M > 0$ such that $|b_n| \leq M$ for all n. Now, for the positive number $\frac{\varepsilon}{2M}$ there exists N_1 such that, for all $n \geq N_1$, $|a_n - L| < \frac{\varepsilon}{2M}$. Similarly, for the positive number $\frac{\varepsilon}{2(|L|+1)}$, there exists N_2 such that, for all $n \geq N_2$, $|b_n - K| < \frac{\varepsilon}{2(|L|+1)}$. Thus for all $n \geq N = \max\{N_1, N_2\}$,

$$|a_n - L||b_n| + |L||b_n - K| < \frac{\varepsilon}{2M} \cdot M + |L| \cdot \frac{\varepsilon}{2(|L|+1)} < \frac{\varepsilon}{2} + \frac{\varepsilon}{2} = \varepsilon.$$

Because, for $n \geq N$, $|a_n b_n - LK| < \varepsilon$, the sequence $\{a_n b_n\}$ is convergent with limit LK. □

Arguments like the above are quite standard in analysis and hence worth understanding thoroughly. For this reason, the next exercises are particularly important.

Exercise 8.11. Prove part (a) of Proposition 8.3. See if you can write a proof that does not treat the case $A = 0$ separately.

Exercise 8.12. Prove part (d) of Proposition 8.3.

Now that we have established Proposition 8.3, we are free to do the sort of limit calculations illustrated in Example 8.3. You should do one or two exercises in which you pay particular attention to precisely which limit laws you use at each step.

Exercise 8.13. Use Proposition 8.3 and the result of Exercise 8.8 to evaluate each limit, indicating at each step which part of the proposition you are using.

(a) $\displaystyle\lim_{n\to\infty} \frac{2}{n+1}$.

(b) $\displaystyle\lim_{n\to\infty} \frac{2n^3 - 3}{n^3 + 1}$.

8.3 Cauchy Sequences

We now know what it means for a sequence a to have limit L. Suppose now we are given a sequence that we suspect has no limit. How could we prove such a claim? We could negate the definition of a limit. We would need to show that for every possible limit L, there exists $\varepsilon > 0$ such that, for all $N \in \mathbb{N}$, there exists $n \geq N$ such that $|a_n - L| \geq \varepsilon$.

Example 8.4. Consider, for example, the sequence given by $a_n = (-1)^n$. This sequence appears to have no limit because all terms with odd index are equal to -1 and all terms with even index are equal to 1. Writing a proof using the negation of the definition of a limit is, however, a bit of a pain. We must begin with an arbitrary L. Suppose first that $L \neq 1$. Thus $|1 - L| > 0$. Take $\varepsilon = |1 - L|$ and let $N \in \mathbb{N}$ be arbitrary. We must find an $n \geq N$ for which $|a_n - L|$ is greater than or equal to this ε. Consider $n = 2N$. Clearly $n \geq N$. Also, $a_n = 1$, and so

$$|a_n - L| = |1 - L| = \varepsilon,$$

proving that L is not the limit. Next consider $L = 1$. Take $\varepsilon = 2$ and let $N \in \mathbb{N}$ be arbitrary. Consider $n = 2N + 1$. Then $a_n = -1$, and so

$$|a_n - L| = |-1 - 1| = 2 = \varepsilon.$$

Thus $L = 1$ is not the limit. We conclude that $\{a_n\}$ does not have a limit.

Although Example 8.4 gives a rather simple divergent sequence, the proof is somewhat awkward. We would like a strong necessary condition for convergence of sequences that gives cleaner proofs that a given sequence is divergent. The condition is the *Cauchy condition.*

Definition 8.2. The sequence a is a **Cauchy sequence** if, given $\varepsilon > 0$, there exists $N \in \mathbb{N}$ such that for all $m, n \geq N$, $|a_n - a_m| < \varepsilon$.

Thus a sequence is Cauchy if, given any error ε, when we go far enough out, any two terms are within ε of one another. The divergent sequence above certainly does not have this property; no matter how far out we go in the sequence, consecutive terms differ by 2. We have indeed found the necessary condition for convergence we seek.

Proposition 8.4. *If the sequence a is convergent, then the sequence is Cauchy.*

Proof. Let $\varepsilon > 0$ be given. We must show that there exists $N \in \mathbb{N}$ such that, for all $m, n \geq N$, $|a_n - a_m| < \varepsilon$.

By hypothesis, a has a limit. Call it L. Thus for the positive number $\frac{\varepsilon}{2}$, there exists M such that for all $n \geq M$, $|a_n - L| < \varepsilon$. We claim we may take $N = M$. Indeed, for all $m, n \geq M$,

$$\begin{aligned} |a_n - a_m| &= |(a_n - L) + (L - a_m)| \\ &\leq |a_n - L| + |L - a_m| \qquad \text{(triangle inequality)} \\ &< \frac{\varepsilon}{2} + \frac{\varepsilon}{2} = \varepsilon. \end{aligned}$$

Thus $\{a_n\}$ is Cauchy. □

8.3.1 Showing that a sequence is Cauchy

You may wonder why we would want to show that a sequence is Cauchy; we know that it is a *necessary* condition for convergence, but we have not determined whether it is *sufficient.* At this point, proving that a sequence is in fact Cauchy only tells us that it has cleared this first hurdle. In the next chapter we will see that, within the real number system, the Cauchy condition is in fact sufficient for convergence. It is of great theoretical importance because it gives a characterization of convergent sequences of real numbers. It is, however, also of practical significance, for it will allow us to determine whether a sequence has a limit even if we have no idea what that limit might be. This situation is much more common than one might guess from looking at exercises in calculus and elementary analysis texts. We therefore give an example and some exercises in which we prove directly that a given sequence is Cauchy.

Example 8.5. The sequence given by $a_n = \frac{1}{n^2}$ is Cauchy.

Proof. Let $\varepsilon > 0$ be given. We must show that there exists $N \in \mathbb{N}$ such that, for all $m, n \geq N$, $|a_n - a_m| < \varepsilon$. Assume without loss of generality that $m > n$. Then

$$\begin{aligned} |a_n - a_m| &= \left| \frac{1}{n^2} - \frac{1}{m^2} \right| \\ &= \frac{1}{n^2} - \frac{1}{m^2} \\ &< \frac{1}{n^2}. \end{aligned}$$

If we take N to be any natural number greater than $\frac{1}{\sqrt{\varepsilon}}$, then for $m > n \geq N$,

$$|a_n - a_m| < \frac{1}{n^2} \leq \frac{1}{N^2} < \varepsilon.$$

□

Exercise 8.14. Prove directly from the definition that the sequence given by $a_n = \frac{3(-1)^n}{n}$ is Cauchy.

Exercise 8.15. Prove directly from the definition that the sequence given by $a_n = \frac{n}{n+1}$ is Cauchy.

8.3.2 Showing that a sequence is divergent

We have shown that if sequence a has a limit, then it is Cauchy. The contrapositive of this statement gives an important tool for showing that a sequence is divergent. Even though it is logically equivalent to Proposition 8.4, we state it here for emphasis.

If a is not Cauchy, then a does not have a limit.

In order to use this result, we must begin by carefully negating the definition of a Cauchy sequence.

Definition 8.3. The sequence a is **not Cauchy** if there exists $\varepsilon > 0$ such that, for all $N \in \mathbb{N}$, there exist $m, n \geq N$ with $|a_n - a_m| \geq \varepsilon$.

Example 8.6. Let us revisit Example 8.4 and prove again that the sequence given by $a_n = (-1)^n$ is divergent. This time we show that this sequence is not Cauchy. Consider $\varepsilon = 1$ and let N be arbitrary. Consider $m = 2N + 1$ and $n = 2N$. Then $m, n \geq N$ and

$$|a_n - a_m| = |(-1)^{2N} - (-1)^{2N+1}| = |1 + 1| = 2 \geq \varepsilon.$$

The sequence is not Cauchy and thus has no limit.

Exercise 8.16. Prove directly from the definition that the sequence given by $a_n = n$ is not Cauchy.

Exercise 8.17. Prove that the sequence given by $a_n = \frac{(-1)^n n}{2n+1}$ considered earlier in the chapter is divergent.

8.3.3 Properties of Cauchy sequences

We collect here a number of propositions about Cauchy sequences of real numbers that are analogous to propositions about convergent sequences of real numbers. It is not surprising that such analogues exist because, in the real number system, a sequence is convergent if and only if it is Cauchy (see the next chapter). However, we can prove these results directly from the definition of a Cauchy sequence, and we wish to do so because we will use several of these results when we discuss the *construction* of the real number system from the rational number system.

Proposition 8.5. *If the sequence a is Cauchy, then a is bounded.*

Proof. This proof is nearly identical to the proof of Proposition 8.2.

Because a is Cauchy, for $\varepsilon = 1$, there exists N such that, for all $n, m \geq N$, $|a_n - a_m| < 1$. In particular, for all $n \geq N$, $|a_n - a_N| \leq 1$. By the reverse triangle inequality, for all $n \geq N$, $|a_n| \leq |a_N| + 1$. If we set $M = \max\{|a_0|, \ldots, |a_{N-1}|, |a_N| + 1\}$, then for all n, $|a_n| \leq M$. □

Proposition 8.6. *Let $a = \{a_n\}$ and $b = \{b_n\}$ be Cauchy and let A be a constant.*

(a) $\{Aa_n\}$ is Cauchy.

(b) $\{a_n + b_n\}$ is Cauchy.

(c) $\{a_n b_n\}$ is Cauchy.

(d) Suppose there exists $\delta > 0$ such that $|a_n| \geq \delta$ for all n. Then $\left\{\frac{1}{a_n}\right\}$ is Cauchy.

Proof. We prove only part (d), leaving the proofs of the other parts to the reader.

Let $\varepsilon > 0$ be given. Because a is Cauchy, for the positive number $\varepsilon\delta^2$, there exists N such that, for all $m, n \geq N$, $|a_n - a_m| < \varepsilon\delta^2$. Then for $m, n \geq N$,

$$\begin{aligned} \left|\frac{1}{a_m} - \frac{1}{a_n}\right| &= \left|\frac{a_n - a_m}{a_m a_n}\right| \\ &= \frac{1}{|a_m||a_n|}|a_n - a_m| \\ &< \frac{1}{\delta^2}\varepsilon\delta^2 = \varepsilon. \end{aligned}$$

Thus $\left\{\frac{1}{a_n}\right\}$ is Cauchy. □

Exercise 8.18. Prove parts (a)–(c) of Proposition 8.6.

8.4 Problems

1. Determine (with justification) whether each of the following sequences is convergent or divergent. If the sequence is convergent, find its limit.

   (a) $a_n = \dfrac{(-1)^n n}{2n^2 + 1}$.

   (b) $a_n = \dfrac{n^2}{n + 1}$.

   (c) $a_n = \dfrac{4n^2}{n^2 + 2n + 1}$.

   (d) The sequence $\{a_n\}$ whose n-th term is the remainder when n is divided by 3.

2. Give an example of each of the following or explain why no such example exists. Prove that the examples you give have the desired properties.

   (a) An increasing sequence that is convergent.

   (b) A convergent sequence with infinitely many positive and infinitely many negative terms.

   (c) A divergent sequence with infinitely many terms equal to zero.

   (d) A sequence of positive terms with limit -0.0001.

   (e) A non-constant sequence with limit 10.

3. Let $a = \{a_n\}$ be a sequence of real numbers. We say that a tends to ∞ and write $\lim_{n\to\infty} a_n = \infty$ if, given M, there exists $N \in \mathbb{N}$ such that, for all $n \geq N$, $a_n > M$.

   (a) Show that $\lim_{n\to\infty} 2n + 3 = \infty$.

(b) Formulate an analogous definition of a sequence tending to $-\infty$.

(c) Prove or give a counterexample: If the sequence a is not bounded, then either $\lim_{n\to\infty} a_n = \infty$ or $\lim_{n\to\infty} a_n = -\infty$.

4. Let r be a real number. Consider the geometric sequence $g = \{r^n\}$. We wish to explore how the convergence or divergence of the sequence depends on the value of r.

   (a) For which (if any) values of r is the sequence g bounded?

   (b) For which (if any) values of r is g convergent? For such r, what is the limit?

   (c) For which (if any) values of r is $\lim_{n\to\infty} r^n = \infty$? For which (if any) values of r is $\lim_{n\to\infty} r^n = -\infty$? (See the previous problem for definitions of these notations.)

5. Let p be a non-constant polynomial. Define a sequence $\{p(n)\}$ by evaluating the polynomial at each non-negative integer. What can you say about $\lim_{n\to\infty} p(n)$?

6. Let a and b be sequences of real numbers. Define a sequence c by setting $c_n = a_n b_n$. We proved that if a and b are both convergent, then so is c. In this problem, you will explore the convergence or divergence of c under other hypotheses on the sequences a and b.

   (a) Suppose $\lim_{n\to\infty} a_n = 0$. Must c be convergent? (Either prove that it must be or give a counterexample.)

   (b) Suppose $\lim_{n\to\infty} a_n = 0$. Could c be convergent even if b is unbounded? Could c be convergent with limit unequal to zero?

   (c) Could c be convergent even if both a and b are divergent?

7. Part (d) of Proposition 8.3 implies that if $\lim_{n\to\infty} a_n = L$ and $L \neq 0$, then $\left\{\frac{1}{a_n}\right\}$ is convergent with limit $\frac{1}{L}$. Consider the converse. In other words, if $\left\{\frac{1}{a_n}\right\}$ is convergent, must $\{a_n\}$ be?

8. Just as the absolute value of a real number measures its distance from 0, we can define the absolute value or *modulus* of a complex number as its distance from the origin of the complex plane. Thus if $z = x + iy$ for $x, y \in \mathbb{R}$, we define
$$|z| = \sqrt{x^2 + y^2}.$$

   (a) For any complex number $z = x + iy$, we define its *complex conjugate* $\bar{z}$ by $\bar{z} = x - iy$. Show that, for any $z \in \mathbb{C}$, $|z|^2 = z\bar{z}$.

   (b) Let $z = x + iy$. Find simple expressions for $z + \bar{z}$ and $z - \bar{z}$.

   (c) (triangle inequality for $\mathbb{C}$) Prove that, for all $z, w \in \mathbb{C}$, $|z + w| \leq |z| + |w|$.

9. Let $\{z_n\}$ be a sequence of complex numbers.

   (a) Using Definition 8.1 as a model, write a definition of the limit of the sequence $\{z_n\}$.

   (b) Consider the sequence of complex numbers with n-th term $z_n = \frac{1+i}{n}$. Does $\lim_{n\to\infty} z_n$ exist?

10. Let $\{z_n\}$ be a sequence of complex numbers. For each n, write $z_n = x_n + iy_n$. Either prove each statement or give a counterexample.

    (a) If $\{z_n\}$ converges, so do $\{x_n\}$ and $\{y_n\}$.

    (b) If $\{x_n\}$ and $\{y_n\}$ converge, so does $\{z_n\}$.

Programming Project: Ratios of Fibonacci Numbers. Consider the Fibonacci sequence $\{F_n\}$ defined recursively by

$$F_0 = 0, F_1 = 1, \text{ and } F_n = F_{n-1} + F_{n-2} \text{ for all } n \geq 2. \tag{8.1}$$

1. Write code to generate the first 20 terms of the sequence.

2. Let $r_n = \frac{F_n}{F_{n-1}}$ for $n \geq 1$. Write code to generate $r_1, \ldots, r_{19}$. Plot this sequence $\{r_n\}$ of ratios.

3. Does the graph suggest that $\{r_n\}$ is monotone (non-increasing or non-decreasing)? Eventually monotone? Convergent? With what limit?

4. In the next chapter we will show that $\{r_n\}$ is convergent. Assume this, and find $\lim_{n\to\infty} r_n$. (Suggestion: Obtain an equation relating r_n and r_{n-1}.)

5. In Chapter 5 we discussed linear recurrence relations such as the above. In Exercise 5.13, you were asked to solve the recurrence defining the Fibonacci sequence. The answer is known as *Binet's formula* and says

$$F_n = \left(\frac{1+\sqrt{5}}{2}\right)^n + \left(\frac{1-\sqrt{5}}{2}\right)^n. \tag{8.2}$$

   Use this explicit formula to find $\lim_{n\to\infty} r_n$.

Chapter 9

A Closer Look at the Real Number System

A book at this level could treat the real number system in two ways, either as a number system already known to us possessing certain familiar properties, or as something to be constructed rigorously from the rational number system. In this chapter, we essentially do both; in the first section, we take the former perspective. Such a perspective may be the right one for a first proofs course. For those wanting a more in-depth real analysis course, we include a section on the rigorous construction of the real numbers as equivalence classes of Cauchy sequences of rational numbers.

9.1 $\mathbb{R}$ as a Complete Ordered Field

9.1.1 Completeness

Both the rational and real number systems form ordered fields. The property of the real number system making it the proper place to do calculus is *completeness*.

The Completeness Axiom for $\mathbb{R}$. Every Cauchy sequence of real numbers converges to a real number.

Recall that we ended the last chapter by showing that if a is a sequence of rational or real numbers and if a converges, then a is Cauchy. Thus the Completeness Axiom states that, in the real number system, the converse is true. Below we will show that the rational number system does not have this property. First, however, we explore some consequences of the completeness of $\mathbb{R}$.

Definition 9.1. Let A be a non-empty subset of $\mathbb{R}$. Then A is **bounded above** if there exists $y \in \mathbb{R}$ such that $x \leq y$ for all $x \in A$. We call y an **upper bound** for A.

Definition 9.2. Let A be a non-empty subset of $\mathbb{R}$. A real number y is the **supremum** or **least upper bound** of A if the following properties hold:

(i) y is an upper bound for A.

(ii) For any upper bound z of A, $y \leq z$.

We denote the supremum of A by $\sup A$. If A is not bounded above, we write $\sup A = \infty$.

Note that in order to show that a number is the supremum of a set, we must show two things. First we must show that the number is an upper bound. Second, we must show that the number is less than or equal to any other upper bound. Sometimes to show the second we will consider the contrapositive; to show that an upper bound y is in fact the supremum, we can show that if $z < y$, then z is not an upper bound.

Example 9.1. Consider $A = \{x \in \mathbb{R} : x < 1/2\}$. Then A is bounded above in $\mathbb{R}$. The numbers 1, 17, 1000, and 1/2 are all examples of upper bounds. We claim that $\sup A = \frac{1}{2}$. To prove this claim we need only show that if $z < \frac{1}{2}$, then z is not an upper bound for A. Consider the real number x defined by $x = \frac{z+\frac{1}{2}}{2}$. Then $z < x < \frac{1}{2}$. Because x is an element of A greater than z, z is not an upper bound for A. The claim is proved.

In the last example, it is not enough to know that if $z < \frac{1}{2}$, then $z \in A$, for it is possible for an upper bound of a set to be an element of the set. For example, if $B = \{x \in \mathbb{R} : x \leq 1\}$, 1 is both an upper bound for B and an element of B.

Example 9.2. Consider $\mathbb{N} \subseteq \mathbb{R}$. Because $\mathbb{N}$ is not bounded above in $\mathbb{R}$, $\sup \mathbb{N} = \infty$.

Exercise 9.1. Write analogous definitions for a **lower bound** and the **infimum** (or **greatest lower bound**) of a non-empty subset A of $\mathbb{R}$. Then give an example of a subset of $\mathbb{R}$ that is bounded below in $\mathbb{R}$ and an example of a subset of $\mathbb{R}$ that is not bounded below in $\mathbb{R}$. We denote the infimum of a set A by $\inf A$.

Exercise 9.2. Let A be a non-empty subset of $\mathbb{R}$, and suppose that y is both an upper bound for A and an element of A. Prove that $y = \sup A$.

If a set has a supremum, this number is clearly unique. An incredibly important consequence of the Completeness Axiom for $\mathbb{R}$ is the *existence* of the supremum of a non-empty set of real numbers that is bounded above. Moreover, our proof of this result illustrates an idea we will use frequently.

Theorem 9.1 (Existence of the supremum)**.** *Let A be a non-empty subset of $\mathbb{R}$ that is bounded above. Then there exists a unique real number* $\sup A$ *that is its supremum.*

The proof of this theorem requires a pair of easy lemmas about limits that we have not yet proved. We will use these lemmas surprisingly often.

Lemma 9.1. *Let c be a convergent sequence with limit L. Suppose that for all n, $c_n \geq 0$. Then $L \geq 0$.*

Proof. Suppose, for a contradiction, that $L < 0$. Let ε be the positive number $\frac{|L|}{2}$. Because c converges to L, there exists $N \in \mathbb{N}$ such that, for all $n \geq N$, $|c_n - L| < \varepsilon$. But then, for all $n \geq N$, $c_n - L < \frac{|L|}{2}$, or, equivalently,

$$c_n < L + \frac{|L|}{2} = \frac{L}{2} < 0.$$

We have reached a contradiction and so we conclude that $L \geq 0$. □

Lemma 9.2 (Non-strict inequalities are preserved in the limit). *Let a and b be convergent sequences with $\lim_{n\to\infty} a_n = \alpha$ and $\lim_{n\to\infty} b_n = \beta$. Suppose there exists N such that for all $n \geq N$, $a_n \leq b_n$. Then $\alpha \leq \beta$.*

Proof. Apply Lemma 9.1 to the sequence $\{b_n - a_n\}$. □

We will often apply this last lemma when one of the sequences is a constant sequence. Thus the lemma has as an immediate consequence that if β is a real number and $a_n \geq \beta$ for all n, then $\lim_{n\to\infty} a_n \geq \beta$.

Proof of Theorem 9.1. Here is the simple idea: We will define two sequences a and b. All terms in a will be elements of A and all terms of b will be upper bounds for A. The sequence a will be non-decreasing, the sequence b will be non-increasing, and the differences $b_n - a_n$ will tend to zero. We will prove that both sequences are Cauchy, hence convergent. Because the differences tend to zero, it will follow that both sequences have the same limit, and this limit will be the supremum for A. If you want to see this idea in action before reading the details of the proof, see Example 9.4.

Because A is non-empty, there exists $a_0 \in A$. Because A is bounded above, there exists $b_0 \in \mathbb{R}$, an upper bound for A. Let $d = b_0 - a_0$. If b_0 is in A, $b_0 = \sup A$ and we are finished. Thus we suppose $d > 0$. Proceed inductively; for $n \geq 1$, having defined a_k and b_k for $0 \leq k \leq n-1$, we define a_n and b_n as follows. Consider $x = \frac{a_{n-1}+b_{n-1}}{2}$. If x is an upper bound for A, set $a_n = a_{n-1}$ and $b_n = x$ and note that $b_n - a_n = \frac{1}{2}(b_{n-1} - a_{n-1})$. Otherwise there exists $a_n \in A$ with $x < a_n \leq b_{n-1}$. In this case, set $b_n = b_{n-1}$ and note that $b_n - a_n < \frac{1}{2}(b_{n-1} - a_{n-1})$.

We now have two sequences a and b with the following properties:

(a) $a_{n-1} \leq a_n$ for all $n \in \mathbb{N}$ and $b_n \leq b_{n-1}$ for all $n \in \mathbb{N}$.

(b) $|b_n - a_n| \leq \frac{1}{2^n} d$ for all $n \geq 0$ and thus $\lim_{n\to\infty}(b_n - a_n) = 0$.

We claim that both sequences are Cauchy. We give the proof for the sequence a and leave the nearly identical proof for sequence b to the reader. Let $\varepsilon > 0$ be given. Assume without loss of generality that $m > n$. Then

$$\begin{aligned} |a_m - a_n| &= a_m - a_n \\ &\leq b_m - a_n \\ &\leq b_n - a_n \\ &\leq \frac{1}{2^n} d. \end{aligned}$$

Suppose N is a natural number for which $\frac{1}{2^N} d < \varepsilon$. Then for all $m > n \geq N$,

$$|a_m - a_n| \leq \frac{1}{2^n} d \leq \frac{1}{2^N} d < \varepsilon$$

and the sequence a is indeed Cauchy.

Because $\mathbb{R}$ is complete, a converges to a real limit α. Similarly, b converges to a real limit β. We claim $\alpha = \beta$. Indeed,

$$\beta - \alpha = \lim_{n\to\infty} b_n - \lim_{n\to\infty} a_n = \lim_{n\to\infty} (b_n - a_n) = 0.$$

We call this common limit α and claim that $\alpha = \sup A$. Fix $x \in A$. Because each b_n is an upper bound for A, $x \leq b_n$ for all n. By Lemma 9.2, $x \leq \lim_{n\to\infty} b_n = \alpha$. Because x was arbitrary, α is an upper bound for A. Now suppose z is an arbitrary upper bound for A. Because every term of the sequence a is an element of the set A, for all n, $a_n \leq z$. Applying Lemma 9.2 again gives $\alpha = \lim_{n\to\infty} a_n \leq z$. Thus α is the least upper bound for A. □

As our first application of Theorem 9.1, we prove the existence of square roots of positive real numbers. We have used this result implicitly in previous chapters, so it will be reassuring to see its proof.

Definition 9.3. Let $c \in \mathbb{R}$. A **real square root** of c is a real number α such that $\alpha^2 = c$.

Proposition 9.1. *Every positive real number c has a positive real square root, denoted $\sqrt{c}$.*

Proof. Let $A = \{x \in \mathbb{R} : x^2 < c\}$. Note that $0 \in A$ and so A is non-empty. We must show that A is bounded above. We claim that any positive y with $c \leq y^2$ is an upper bound for A. If not, for some such y we could find $x \in A$ with $0 < y < x$. But then $y^2 < x^2 < c$, a contradiction. Our claim is established. To find an upper bound for A, we consider two cases. In the first case, if $1 \leq c$, then $c \leq c^2$ and c itself is an upper bound. In the second case, if $c \leq 1$, then $c \leq 1^2$ and 1 is an upper bound.

Because A is a non-empty subset of $\mathbb{R}$ that is bounded above, by Theorem 9.1, there exists a real number α that is the supremum of A. We claim that $\alpha^2 = c$. From the proof of Theorem 9.1, we know that α is the common limit of a sequence a of elements of A and a sequence b of upper bounds for A. Because the terms of a are in A, for all n, $a_n^2 < c$. By Lemma 9.2, $\alpha^2 = \lim_{n\to\infty} a_n^2 \leq c$. On the other hand, because the terms of b are upper bounds for A, for all n, $c \leq b_n^2$. Again by Lemma 9.2, $c \leq \lim_{n\to\infty} b_n^2 = \alpha^2$. Thus $\alpha^2 = c$. □

The proof of Theorem 9.1 uses rather heavily the fact that the sequence a is non-decreasing and the sequence b is non-increasing. We call a sequence *monotone* if it is either non-decreasing or non-increasing. The following theorem on monotone sequences is an easy consequence of Theorem 9.1. Nevertheless, it provides a very useful sufficient condition for convergence of monotone sequences.

Theorem 9.2 (Monotone Sequence Theorem). *Let b be a sequence of real numbers.*

(a) If a is non-decreasing and bounded above, then a converges.

(b) If a is non-increasing and bounded below, then a converges.

Proof. We prove part (a); the proof of part (b) is similar and is left as an exercise.

Let A be the set of real numbers appearing as terms in the sequence a, i.e., $A = \{a_n : n \geq 0\}$. Because the sequence a is bounded above, A is a (non-empty) subset of $\mathbb{R}$ that is bounded above. Thus by Theorem 9.1, there is a real number $\sup A$ that is the supremum of the set A. We claim that the sequence a converges to $\sup A$. Let $\varepsilon > 0$ be given. Consider the real number $\sup A - \varepsilon$. Because $\sup A$ is the least upper bound for A, some element of A exceeds $\sup A - \varepsilon$. Thus there exists N such that $\sup A - \varepsilon < a_N \leq \sup A$. Because a is non-decreasing and bounded above by $\sup A$, for all $n \geq N$,

$$|a_n - \sup A| = \sup A - a_n \leq \sup A - a_N < \varepsilon.$$

Thus $\lim_{n\to\infty} a_n = \sup A$. □

Note the distinction between the sequence a and the set A in the proof of Theorem 9.2. The sequence a is a function on $\mathbb{N}_0$. We understand it as we do any function, by looking at its values a_n for elements n of the domain. *The set A is the image of $\mathbb{N}_0$ under the function a.* We have lost all information about which element of the domain produces which elements of A. An example may further illustrate the distinction; consider the sequence $a = \{(-1)^n\}$. We sometimes write this sequence as the ordered list $\{1, -1, 1, -1, 1, -1, \ldots\}$. As with all sequences, the sequence a has infinitely many terms. In this case, the image of the sequence is $A = \{-1, 1\}$, which is a finite set.

Exercise 9.3. Give an example of two sequences that are not equal but that have the same image set.

Exercise 9.4. Prove part (b) of the Monotone Sequence Theorem.

We will use the Monotone Sequence Theorem often. It is useful both in the proofs of other theorems and to show that specific sequences have limits. It is especially useful when numerical exploration suggests to us that a sequence is monotone and convergent but when it is unclear what the limit might be or when we do not have an explicit formula for the sequence. We consider such an example.

Figure 9.1: First 20 terms of the sequence a from Example 9.3.

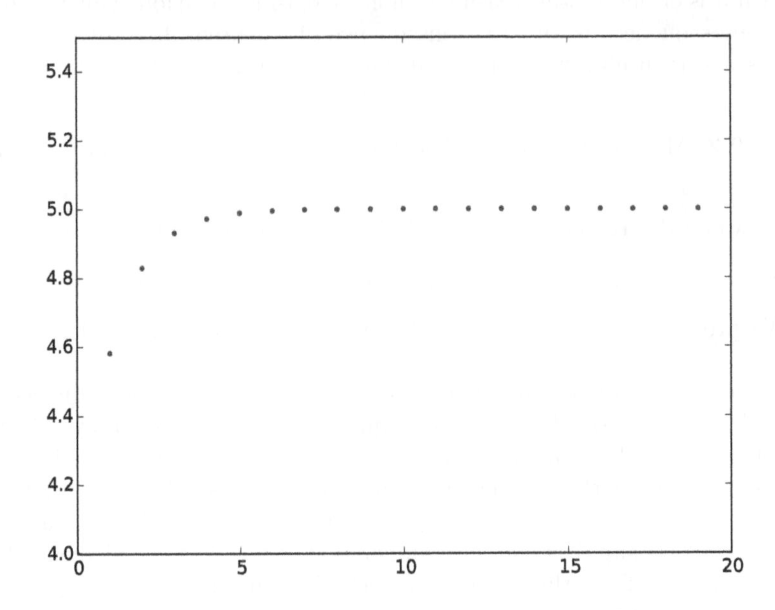

Example 9.3. Define a sequence a recursively by setting $a_0 = 4$ and $a_n = \sqrt{4a_{n-1} + 5}$ for $n \geq 1$. It is trivial to generate the first 20 terms with a few lines of code.

```
N=20 #number of terms to generate
a=[4.0] #initial condition

for n in range(1,N):
    a.append(sqrt(4.0*a[n-1]+5.0)) #recursive formula

plot(a,".")
show()
```

Figure 9.1 shows the plot. We conjecture that a is non-decreasing and convergent, perhaps with a limit of 5. The proof of the conjecture requires three steps. First we will prove that a is bounded above by 5 and below by 0. Second, we will prove that a is non-decreasing. The Monotone Sequence Theorem will then imply that a is convergent. Third, we will solve for the limit algebraically.

Our proof that a is bounded above by 5 is by induction on $n \in \mathbb{N}_0$. Because $a_0 = 4$, the result holds for $n = 0$. Next, suppose the result holds for some k and consider $k + 1$. By the recursive definition for a and the inductive hypothesis,

$$a_{k+1} = \sqrt{4a_k + 5} \leq \sqrt{4(5) + 5} = 5,$$

and so the result holds for $k+1$. By the principle of mathematical induction, it holds for all $n \in \mathbb{N}_0$. The proof that a is bounded below by 0 is similar and is omitted.

In order to show that a is non-decreasing, we note that because $a_n - 5 \leq 0$ and $a_n + 1 \geq 0$, $a_n^2 - 4a_n - 5 = (a_n - 5)(a_n + 1) \leq 0$. Thus for all n, $a_n^2 \leq 4a_n + 5 = a_{n+1}^2$. Because all terms of a are non-negative, we may conclude that $a_n \leq a_{n+1}$ for all n.

Because a is non-decreasing and bounded above, by the Monotone Sequence Theorem, a converges. Call its limit L. Taking the limit of both sides of $a_n^2 - 4a_n - 5 = 0$ and using properties of limits shows that $L^2 - 4L - 5 = 0$. This quadratic has solutions $L = 5$ and $L = -1$, but only 5 can be the limit of a because a consists of non-negative terms. We conclude that $\lim_{n\to\infty} a_n = 5$.

Exercise 9.5. In Problem 9 of Chapter 2, we considered the boundedness and monotonicity of the sequence a defined recursively by $a_0 = 1$ and $a_n = \sqrt{2a_{n-1} + 3}$. Do that problem if you have not already done so. Determine whether a is convergent. If it is, find its limit.

9.1.2 Why $\mathbb{Q}$ is not complete

In this section, let

$$A = \{r \in \mathbb{Q} : r^2 < 2\}.$$

The set A is non-empty because $1 \in A$. It is also bounded above in $\mathbb{Q}$. Indeed, as we argued in the proof of Proposition 9.1, any positive y with $2 \leq y^2$ is an upper bound for A. In particular, because $2 \leq 2^2$, 2 is an upper bound for A in $\mathbb{Q}$.

We may also view A as a subset of $\mathbb{R}$, and, just as above, we conclude that A is a non-empty subset of $\mathbb{R}$ that is bounded above. By Theorem 9.1, there exists a real number that is the supremum of A. Let $\alpha = \sup A$. An identical argument to the one given in the proof of Proposition 9.1 shows that $\alpha^2 = 2$.

We show that α is not rational. We give a proof by contradiction. Thus we suppose α is a positive real number such that $\alpha^2 = 2$ and that there exist integers m and n with $\alpha = \frac{m}{n}$. We may assume m and n are positive and that they have no common factors (for otherwise we could cancel any common factors in the ratio $\frac{m}{n}$ and still have a representation of α as a ratio of integers). Now, because $\alpha^2 = 2$,

$$2 = \alpha^2 = \frac{m^2}{n^2} \iff m^2 = 2n^2. \tag{9.1}$$

Thus 2 divides $m^2 = m \cdot m$. Because 2 is prime and 2 divides the product $m \cdot m$, 2 must divide one of the factors, i.e., 2 divides m. Thus there exists $k \in \mathbb{N}$ such that $m = 2k$. Substituting into (9.1) gives

$$4k^2 = 2n^2 \iff n^2 = 2k^2.$$

By the same reasoning as above, 2 divides n. We conclude that both m and n are divisible by 2, contradicting our assumption that m and n have no common factors. We conclude that there do not exist integers m and n such that $\alpha = \frac{m}{n}$.

We may now see that $\mathbb{Q}$ is not complete, for if it were, the argument given to prove Theorem 9.1 would prove that every non-empty subset of $\mathbb{Q}$ that is bounded above in $\mathbb{Q}$ has a least upper bound *in* $\mathbb{Q}$. However, our set A is an example of a non-empty subset of $\mathbb{Q}$ that is bounded above in $\mathbb{Q}$ but does not have this property. If we prefer, we can think of this phenomenon in terms of sequences instead; using the method from the proof of Theorem 9.1, we can obtain a sequence a in A converging to $\sup A$. Because this sequence is convergent, it is Cauchy. Therefore, we have found a Cauchy sequence of rational numbers whose limit is not a rational number.

9.1.3 Algorithms for approximating $\sqrt{2}$

In the last subsection, we saw that there exist sequences of rational numbers converging to the real number $\sqrt{2}$. We did not, however, discuss in detail how to obtain any specific sequences with limit $\sqrt{2}$. We do so in this subsection. We are not so much interested in approximating $\sqrt{2}$ as we are in seeing the ideas from Theorem 9.1 and the Monotone Sequence Theorem in action. As we will see many times throughout this book, the right kind of proof of a theorem can be used to generate an algorithm that we can implement on a computer.

Throughout this subsection, let

$$A = \{x \in \mathbb{R} : x^2 < 2\}.$$

Then $\sup A = \sqrt{2}$. Our first algorithm to approximate $\sqrt{2}$ uses the idea from the proof of Theorem 9.1.

Example 9.4. We inductively construct two sequences a and b of rational numbers. Let $a_0 = 1$. Because $1^2 = 1 < 2$, a_0 is in A. Let $b_0 = 2$. As we noted above, b_0 is an upper bound for A. Proceed inductively using precisely the method from the proof of Theorem 9.1. Both sequences have limit $\sup A = \sqrt{2}$. Observe that $a_n \leq \sqrt{2} \leq b_n$ for all n and that $|b_n - a_n| \leq 2^{-n}|b_0 - a_0| = 2^{-n}$. Thus

$$|a_n - \sqrt{2}| = \sqrt{2} - a_n \leq b_n - a_n \leq 2^{-n}$$
$$|b_n - \sqrt{2}| = b_n - \sqrt{2} \leq b_n - a_n \leq 2^{-n}.$$

We have not only obtained two explicit sequences approximating $\sqrt{2}$, but also have an upper bound on the error of the approximation, namely, $|a_n - \sqrt{2}| \leq 2^{-n}$.

Let's look at a program that uses this algorithm to approximate $\sqrt{2}$.

```
# a and b are sequences of rational numbers
# Initial conditions. Use floats to ensure floating point
    division
```

```
a = [1.0]
b = [2.0]

# loop a maximum of 1000 times. In practice, <100 iterations are
    required to exceed the precision of our basic floating point
    variable
for n in range(1,1000):
    x = (a[n-1] + b[n-1]) / 2.0

    if x**2 < 2: #x is in A
        # add x to a. b[n] is unchanged from b[n-1]
        a.append( x )
        b.append( b[n-1] )
    else: # x is an upper bound for A
        # add x to b. a[n] is unchanged from a[n-1]
        a.append( a[n-1] )
        b.append( x )

    # check that something has changed this iteration. If it
        hasn't, we've exceeded the precision of our floating
        point variables
    if a[n] == a[n-1] and b[n] == b[n-1]:
        print("Floating point precision exceeded.")
        break

    # may never happen, but it would be a stopping point if it
        did
    if a[n] == b[n]:
        print("a[n] = b[n]. Limit reached.")
        break

# approximation complete, output results
print("Stopped after "+str(n)+" iterations.")
print("Max a[n] = "+ str(a[n]))
print("Min b[n] = "+ str(b[n]))
```

We obtain the following output:

```
Floating point precision exceeded.
Stopped after 53 iterations.
Max a[n] = 1.41421356237
Min b[n] = 1.41421356237
```

We have thus obtained 1.41421356237 as our approximation to $\sqrt{2}$.

Our second algorithm also generates a sequence converging to $\sqrt{2}$, though the convergence is faster.

Example 9.5. Define a sequence a recursively by setting $a_0 = 1$ and

$$a_n = \frac{a_{n-1} + \frac{2}{a_{n-1}}}{2}, \quad n \geq 1. \tag{9.2}$$

Clearly all terms of the sequence a are non-negative. Writing the code to generate and plot this sequence is easy and is left as the next exercise. The graphical and numerical evidence suggests that this sequence is non-increasing for $n \geq 1$ and bounded below. We prove these claims and then show that the sequence in fact has limit $\sqrt{2}$.

We show first that $a_n \geq \sqrt{2}$ for $n \geq 1$. Indeed, by the arithmetic-geometric mean (AGM) inequality for two positive real numbers (see Problem 14 in Chapter 2),

$$a_n = \frac{a_{n-1} + \frac{2}{a_{n-1}}}{2} \geq \sqrt{a_{n-1} \cdot \frac{2}{a_{n-1}}} = \sqrt{2}.$$

Next, we note that $a_n \leq a_{n-1}$ if and only if

$$a_{n-1} + \frac{2}{a_{n-1}} \leq 2a_{n-1},$$

or, equivalently,

$$2 \leq a_{n-1}^2,$$

which is true if $a_{n-1} \geq \sqrt{2}$. Thus after the first term, the sequence is indeed non-increasing. By the Monotone Sequence Theorem, a is convergent to some real number L. Taking the limit of both sides of (9.2) and using Proposition 8.3 gives

$$L = \frac{L + \frac{2}{L}}{2}.$$

Solving for L gives $L = \pm\sqrt{2}$. Because all terms of a are non-negative, a has limit $\sqrt{2}$.

Exercise 9.6. Write code to generate the sequence in (9.2). As in Example 9.4, write your program so that it stops when nothing significant has changed and so that it reports the number of iterations needed to get to this point.

Exercise 9.7. Let α be any positive real number. Modify Example 9.5 to obtain a sequence of real numbers with limit $\sqrt{\alpha}$. Is it possible to use this argument to prove that every positive real number α has a positive square root? Explain.

9.2 Construction of $\mathbb{R}$

$\mathbb{R}$ contains $\mathbb{Q}$ and is in fact the smallest complete ordered field containing $\mathbb{Q}$. Thus one way to think of $\mathbb{R}$ is as $\mathbb{Q}$ together with the limits of all Cauchy sequences of elements of $\mathbb{Q}$. In this section, we flesh out this idea, indicating precisely how to construct $\mathbb{R}$ from $\mathbb{Q}$. Because we will end this section where we started the chapter, this section could be omitted without impacting your ability to read the remainder of the book.

9.2.1 An equivalence relation on Cauchy sequences of rational numbers

Because $\mathbb{Q}$ is not complete, there are Cauchy sequences of rational numbers that do not converge to a rational number. When we encounter such a sequence, we could put a label on it and use this label as our name for the limit of the sequence. For example, for the sequence

$$\{2, 2.7, 2.71, 2.718, 2.7182, \ldots\}$$

we might assign the label e and throw it into the set of real numbers we are constructing. But then what do we do with the sequences

$$\{4, 2.7, 2.71, 2.718, 2.7182, \ldots\}$$
$$\{1, 2, 2.5, 2.\bar{6}, 2.708\bar{3}, 2.71\bar{6}, 2.7180\bar{5}, \ldots\}?$$

These three sequences of rational numbers are clearly not the same sequence, but each seems to be giving progressively better approximations of the same real number. We would therefore like to think of them as equivalent. To capture this idea, we define a relation on the set of all Cauchy sequences of rational numbers.

Definition 9.4. Let a and b be Cauchy sequences of rational numbers. We say that a and b are **equivalent** and write $a \sim b$ if, given $M \in \mathbb{N}$, there exists $N \in \mathbb{N}$ such that, for all $n \geq N$, $|a_n - b_n| < \frac{1}{M}$.

Proposition 9.2. *$\sim$ is an equivalence relation on the set of all Cauchy sequences of rational numbers.*

Proof. We prove only that $\sim$ is transitive; the other two properties are trivial. Let a, b, and c be Cauchy sequences with $a \sim b$ and $b \sim c$. We must show that $a \sim c$. Let $M \in \mathbb{N}$ be given. Because $a \sim b$, there exists $N_1 \in \mathbb{N}$ such that, for all $n \geq N_1$, $|a_n - b_n| < \frac{1}{2M}$. Similarly, because $b \sim c$, there exists $N_2 \in \mathbb{N}$ such that, for all $n \geq N_2$, $|b_n - c_n| < \frac{1}{2M}$. Then for all $n \geq N = \max\{N_1, N_2\}$,

$$\begin{aligned} |a_n - c_n| &= |a_n - b_n + b_n - c_n| \\ &\leq |a_n - b_n| + |b_n - c_n| \\ &< \frac{1}{2M} + \frac{1}{2M} = \frac{1}{M}, \end{aligned}$$

and $a \sim c$. □

Exercise 9.8. Prove that the constant sequence $c = \{1, 1, 1, \ldots\}$ and the sequence $a = \left\{1 - \frac{(-1)^n}{n+1}\right\}$ are equivalent. Then define a third sequence b differing from each of a and c for infinitely many values of n but equivalent to both.

Exercise 9.9. Let a and b be Cauchy sequences of rational numbers. Write down carefully what it means for these two sequences to be **not equivalent**.

We may now define real numbers.

Definition 9.5. A **real number** x is an equivalence class of Cauchy sequences of rational numbers under the equivalence relation $\sim$.

This definition may strike you as odd, but it is no stranger than our definition of a rational number as an equivalence class of $\mathbb{Z}\times(\mathbb{Z}\setminus\{0\})$ under the equivalence relation

$$(p,q)\sim(r,s) \iff ps=qr.$$

Just as each rational number can be represented by any of a number of equivalent ratios of integers, each real number can be represented by any of a number of equivalent Cauchy sequences of rational numbers. For example, the three sequences

$$\{2, 2.7, 2.71, 2.718, 2.7182, \ldots\}$$
$$\{4, 2.7, 2.71, 2.718, 2.7182, \ldots\}$$
$$\{1, 2, 2.5, 2.\bar{6}, 2.708\bar{3}, 2.71\bar{6}, 2.7180\bar{5}, \ldots\}$$

all represent e, and the three sequences in Exercise 9.8 all represent the number 1.

9.2.2 Operations on $\mathbb{R}$

At the moment, we have only defined the elements of the set of real numbers. Because we want the real numbers to form an ordered field, we must define operations of addition and multiplication, we must define an order relation, and we must show that the ordered field axioms are satisfied.

Let x and y be real numbers. How should we define $x+y$? We use the same approach we have used previously to define operations on sets of equivalence classes. (See, for example, the sections on modular arithmetic and rational numbers in Chapter 6.) We take a Cauchy sequence $\{x_n\}$ of rationals from the equivalence class x and a Cauchy sequence $\{y_n\}$ of rationals from the equivalence class y. We consider the sequence $\{x_n+y_n\}$. Because x_n and y_n are rational and there is a well-defined binary operation of addition on $\mathbb{Q}$, $\{x_n+y_n\}$ is a sequence of rational numbers. In the previous chapter, in Proposition 8.6, we proved that such a sum of Cauchy sequences is itself Cauchy. Thus $\{x_n+y_n\}$ is an element of an equivalence class of Cauchy sequences of rationals. We define $x+y$ to be this class.

Is $x+y$ well-defined? In other words, is it independent of the Cauchy sequences $\{x_n\}$ and $\{y_n\}$ we choose to represent x and y? The first part of the next proposition says that it is. The second part says that an analogous statement holds for pointwise products of Cauchy sequences.

Proposition 9.3. *Let $\{x_n\}$, $\{x'_n\}$, $\{y_n\}$, and $\{y'_n\}$ be Cauchy sequences of rational numbers. If $\{x_n\}\sim\{x'_n\}$ and $\{y_n\}\sim\{y'_n\}$, then*

(a) $\{x_n+y_n\}\sim\{x'_n+y'_n\}$.

(b) $\{x_n y_n\} \sim \{x'_n y'_n\}$.

Proof. At this point, both proofs are standard. We prove the harder part (b) and leave part (a) as an exercise.

Let $M \in \mathbb{N}$ be given. We must show that there exists $N \in \mathbb{N}$ such that, for all $n \geq N$, $|x_n y_n - x'_n y'_n| < \frac{1}{M}$. By the triangle inequality,

$$\begin{aligned} |x_n y_n - x'_n y'_n| &= |(x_n y_n - x_n y'_n) + (x_n y'_n - x'_n y'_n)| \\ &\leq |x_n||y_n - y'_n| + |x_n - x'_n||y'_n|. \end{aligned}$$

All four sequences are bounded because they are Cauchy. Thus there exist natural numbers A and B such that $|x_n| \leq A$ and $|y'_n| \leq B$ for all n. Thus for all n,

$$|x_n y_n - x'_n y'_n| \leq A|y_n - y'_n| + B|x_n - x'_n|.$$

Because $\{x_n\} \sim \{x'_n\}$, there exists N_1 such that, for all $n \geq N_1$, $|x_n - x'_n| < \frac{1}{2MB}$. Similarly, there exists N_2 such that, for all $n \geq N_2$, $|y_n - y'_n| < \frac{1}{2MA}$. If we set $N = \max\{N_1, N_2\}$, then for all $n \geq N$,

$$|x_n y_n - x'_n y'_n| < A \cdot \frac{1}{2MA} + B \cdot \frac{1}{2MB} = \frac{1}{M},$$

as desired. □

Exercise 9.10. Prove part (a) of Proposition 9.3.

With the above motivation and proposition, we may give the precise definition of the operations on $\mathbb{R}$ and know that these are indeed well-defined.

Definition 9.6. Let $\mathbb{R}$ be the set of all equivalence classes of Cauchy sequences of rational numbers under the equivalence relation $\sim$. Let $x, y \in \mathbb{R}$, and let $\{x_n\}$ and $\{y_n\}$ be any Cauchy sequences representing x and y, respectively. Then $x + y$ is the equivalence class containing $\{x_n + y_n\}$ and xy is the equivalence class containing $\{x_n y_n\}$.

9.2.3 Verifying the field axioms

We first saw the axioms for a field in Chapter 1, Definition 1.3. We gave the field axioms again in Chapter 6 when we discussed algebraic structures more generally. We do not repeat the field axioms here. Of course, our claim is that $\mathbb{R}$ with the two operations defined above is a field. We will not give a complete proof because, for the most part, the verifications are easy. For example, suppose we want to prove that if $x, y \in \mathbb{R}$, $x + y = y + x$. Let $\{x_n\}$ and $\{y_n\}$ be Cauchy sequences of rational numbers representing x and y, respectively. Then $x + y$ is the equivalence class containing $\{x_n + y_n\}$ whereas $y + x$ is the equivalence class containing $\{y_n + x_n\}$. *But because $x_n, y_n \in \mathbb{Q}$ and $\mathbb{Q}$ is a field, for all n, $x_n + y_n = y_n + x_n$, meaning that $\{x_n + y_n\}$ and $\{y_n + x_n\}$ are the same sequence.* Thus $x + y = y + x$.

Exercise 9.11. Prove the associative property for multiplication in $\mathbb{R}$ using the same kind of argument given in the last paragraph.

The additive identity 0 in $\mathbb{R}$ is the equivalence class containing the sequence $\{0, 0, 0, \ldots\}$, and the multiplicative identity 1 is the equivalence class containing the sequence $\{1, 1, 1, \ldots\}$. The one axiom that is a bit tricky to verify is the one that asserts the existence of a multiplicative inverse x^{-1} for a non-zero real number x. We consider this verification in some detail.

Take a real number $x \neq 0$ and let $\{x_n\}$ be any Cauchy sequence representing x. We would like to define x^{-1} to be the equivalence class associated with the sequence $\{x_n^{-1}\}$. *Before reading the next paragraph*, think about the following exercise:

Exercise 9.12. What issues must be settled before we can define x^{-1} to be the equivalence class containing the sequence $\{x_n^{-1}\}$? (I can think of at least three.)

We require three propositions before we can define x^{-1} in this manner. First, because x_n^{-1} is only defined if $x_n \neq 0$, we must prove that there exists $\{x_n\}$ representing x such that $x_n \neq 0$ for all n. Second, we must prove that, for such a sequence representing x, the sequence $\{x_n^{-1}\}$ is Cauchy. Actually, this statement follows from part (d) of Proposition 8.6 in the previous chapter. Third, we must prove that if $\{x_n\} \sim \{x'_n\}$ (with all terms non-zero), $\{x_n^{-1}\} \sim \{(x'_n)^{-1}\}$.

The next proposition will help us address our first issue. In simple language, it says that, if x is a non-zero real number, any sequence representing x is eventually bounded away from 0. As an immediate consequence, by dropping finitely many terms at the beginning of the sequence if necessary, we may assume that any sequence $\{x_n\}$ representing such an x has only non-zero terms.

Proposition 9.4. *Suppose $x \in \mathbb{R}$ and $x \neq 0$. Then there exists $M \in \mathbb{N}$ such that, if $\{x_n\}$ is any sequence representing x, there exists $N \in \mathbb{N}$ such that, for all $n \geq N$, $|x_n| \geq \frac{1}{M}$.*

Proof. Let $\{x_n\}$ be any Cauchy sequence representing x. Because $x \neq 0$, $\{x_n\}$ is not equivalent to any sequence representing zero. In particular, it is not equivalent to the constant sequence $\{0, 0, 0, \ldots\}$. Negating the definition of equivalence, we see that there exists $K \in \mathbb{N}$ such that, for all $N \in \mathbb{N}$, there exists $n \geq N$ with $|x_n| \geq \frac{1}{K}$.

Now consider the natural number $2K$. Because $\{x_n\}$ is Cauchy, there exists N_1 such that, for all $m, n \geq N_1$, $|x_m - x_n| < \frac{1}{2K}$. Fix $m \geq N_1$ with $|x_m| \geq \frac{1}{K}$. Then for all $n \geq N_1$, by the reverse triangle inequality,

$$|x_n| = |x_n - x_m + x_m| \geq |x_m| - |x_n - x_m| > \frac{1}{K} - \frac{1}{2K} = \frac{1}{2K}.$$

Our proof is not complete, for we must show that there exists a single lower bound for the tail of *any* sequence representing x. Thus let $\{x'_n\}$ be any other sequence representing x. Because it is in the same equivalence class as $\{x_n\}$, given $M = 4K$, there exists N_2 such that, for all $n \geq N_2$, $|x'_n - x_n| < \frac{1}{4K}$. Then for all $n \geq N = \max\{N_1, N_2\}$,

$$|x'_n| = |x'_n - x_n + x_n| \geq |x_n| - |x'_n - x_n| > \frac{1}{2K} - \frac{1}{4K} = \frac{1}{4K} = \frac{1}{M}.$$

□

The next exercise completes our argument that, if $x \neq 0$, x^{-1} as defined above makes sense and is well-defined.

Exercise 9.13. Suppose $\{x_n\} \sim \{x'_n\}$ but $\{x_n\}$ is not equivalent to the zero sequence. (Thus we may assume $x_n, x'_n \neq 0$ for all n.) Prove that $\{x_n^{-1}\} \sim \{(x'_n)^{-1}\}$.

With this verification complete, we are prepared to move on, accepting that the operations of addition and multiplication in $\mathbb{R}$ satisfy all the field axioms.

9.2.4 Defining order

The field axioms capture all the important properties of the algebraic structure of the real number system. We, however, are interested in analysis. We know that analysis is concerned with making the notions of calculus precise. First and foremost, we need to be able to talk about limits. As we have already seen, the definition of a limit requires us to be able to compare real numbers. Thus it is essential to us to have an order relation on $\mathbb{R}$.

If $x \in \mathbb{R}$, what should it mean to say that $x > 0$? Because x is an equivalence class of Cauchy sequences of rational numbers and 0 is the name for the equivalence class containing the sequence $\{0, 0, 0, \ldots\}$, our definition of $x > 0$ must describe a condition on Cauchy sequences. Naively, we might suggest that we say $x > 0$ if x is represented by a Cauchy sequence that is eventually positive. Such a definition will not work, because 0 itself is represented by many sequences of positive rational numbers such as $\{\frac{1}{n}\}$ and $\{2^{-n}\}$. We would then be forced to say that $0 > 0$, and our order relation would not satisfy the trichotomy axiom (OF1). We need a stronger condition. The informal version says that $x > 0$ if every Cauchy sequence representing x is eventually positive *and bounded away from zero.*

Definition 9.7. Let $x \in \mathbb{R}$. We say x is **positive** and write $x > 0$ if there exists $M \in \mathbb{N}$ such that, for every sequence $\{x_n\}$ representing x, there exists $N \in \mathbb{N}$ such that, for all $n \geq N$, $x_n \geq \frac{1}{M}$.

The condition in Definition 9.7 bears a striking resemblance to the condition in Proposition 9.4 satisfied by Cauchy sequences representing non-zero real numbers. In fact, a refinement of Proposition 9.4 is our key step in establishing the ordered field axioms for $\mathbb{R}$. We state this refinement as a lemma.

Lemma 9.3. *Suppose $x \in \mathbb{R}$ and $x \neq 0$. Then one of the following holds:*

(i) There exists $M \in \mathbb{N}$ such that, for all $\{x_n\}$ representing x, there exists $N \in \mathbb{N}$ such that, for all $n \geq N$, $x_n \geq \frac{1}{M}$.

(ii) There exists $M \in \mathbb{N}$ such that, for all $\{x_n\}$ representing x, there exists $N \in \mathbb{N}$ such that, for all $n \geq N$, $-x_n \geq \frac{1}{M}$.

Proof. Proposition 9.4 states that if x is non-zero, every sequence representing x is eventually bounded away from zero. Our refinement states that, in fact, every sequence representing x is either eventually positive and bounded away from zero or eventually negative and bounded away from zero. Furthermore, it states that if one sequence representing x is eventually positive (resp., negative), then every other sequence representing x is eventually positive (resp., negative).

Here are the details. Take $x \neq 0$ and $\{x_n\}$ representing x. By Proposition 9.4, there exist natural numbers M and N_1 such that, for all $n \geq N_1$, $|x_n| \geq \frac{1}{M}$. Because $\{x_n\}$ is Cauchy, there exists $N_2 \in \mathbb{N}$ such that, for all $m, n \geq N_2$, $|x_m - x_n| < \frac{1}{M}$. Let $N = \max\{N_1, N_2\}$. Certainly $|x_N| \geq \frac{1}{M}$. We assume for definiteness that $x_N \geq \frac{1}{M}$. The other case follows similarly. We claim that, in fact, for all $n \geq N$, $x_n \geq \frac{1}{M}$. We already know the inequality is true for $|x_n|$, so we need only show that, for all $n \geq N$, x_n is positive. Observe, for all $n \geq N$,

$$x_n = x_n - x_N + x_N \geq x_N - |x_n - x_N| > \frac{1}{M} - \frac{1}{M} = 0.$$

We have now shown that one sequence representing x satisfies the condition (i). We must show every other sequence representing x does as well. Fix another sequence $\{x'_n\}$ representing x. By Proposition 9.4, for the same M as above, there exists N_3 (not necessarily equal to N_1) such that, for all $n \geq N_3$, $|x'_n| \geq \frac{1}{M}$. Furthermore, because $\{x_n\} \sim \{x'_n\}$, there exists N_4 such that, for all $n \geq N_4$, $|x'_n - x_n| < \frac{1}{M}$. Thus if $K = \max\{N, N_3, N_4\}$ and $n \geq K$,

$$x'_n = x'_n - x_n + x_n \geq x_n - |x_n - x'_n| \geq \frac{1}{M} - \frac{1}{M} = 0,$$

as desired. □

Theorem 9.3. *Let $P = \{x \in \mathbb{R} : x > 0\}$, with $x > 0$ defined as above. Then P has the properties (OF1)–(OF3) given in Definition 1.4.*

Proof. We establish the trichotomy property and leave the verifications of the other two properties as an easy exercise.

Suppose first that $x = 0$. Then $\{0, 0, 0, \ldots\}$ is a sequence representing 0 that does not satisfy the condition of Definition 9.7, and so $0 > 0$ is false. Also, $\{0, 0, 0, \ldots\}$ is a sequence representing -0 that does not satisfy the condition of Definition 9.7, and so $-0 > 0$ is also false. Therefore for the real number 0, precisely one of the three statements is true.

Now suppose x is a non-zero real number. One of the conditions in Lemma 9.3 holds. Suppose it is the first so that, by definition, $x > 0$. (The proof in the other case is identical.) Our proof is not complete, for we must argue that $-x > 0$ is false. Because the order relation on $\mathbb{Q}$ satisfies (OF1), a sequence of rational numbers can not simultaneously satisfy both $x_n \geq \frac{1}{M}$ and $-x_n \geq \frac{1}{M}$. Thus $-x > 0$ is false. $\square$

Exercise 9.14. Prove (OF2) and (OF3). That is, show that if $x, y \in \mathbb{R}$ and $x, y > 0$, then $x + y > 0$ and $xy > 0$.

More generally, for $x, y \in \mathbb{R}$, we write $x < y$ or $y > x$ if and only if $y - x > 0$ and we write $x \leq y$ or $y \geq x$ if and only if $y - x > 0$ or $y - x = 0$. It is not hard to see that $<$ is an order relation on $\mathbb{R}$. (Refer to Definition 4.4 for the properties of an order relation.) The comparability and non-reflexivity properties both follow from trichotomy. Transitivity follows from (OF2). We leave the details as an exercise.

Exercise 9.15. Use Theorem 9.3 to prove that, if $x < y$ and $y < z$, then $x < z$.

In Chapter 1, we showed that many general inequalities such as the triangle inequality and its variants follow from the ordered field axioms, and we have freely used those inequalities in this section to make estimates on rational numbers. Now that we have verified the ordered field axioms for the set of real numbers we have constructed, we can freely use all those inequalities in this context as well. Perhaps surprisingly, establishing an inequality between specific real numbers is harder. So far, we have only talked about how to prove that a real number is positive. How do we establish other inequalities between real numbers? Just as positivity of a real number x is demonstrated by verifying that a Cauchy sequence representing x satisfies some condition, an inequality between x and y can be established by proving an inequality relating the terms of a pair of Cauchy sequences representing them. This next proposition will play a significant role in the next section when we consider limits and completeness.

Proposition 9.5. *Let x be a real number represented by $\{x_n\}$ and let y be a real number represented by $\{y_n\}$. If there exists $N \in \mathbb{N}$ such that $x_n \leq y_n$ for all $n \geq N$, then $x \leq y$.*

Proof. Suppose not. Then $x - y > 0$. Now, $\{x_n - y_n\}$ is a Cauchy sequence representing $x - y$. Because $x - y > 0$, there exist natural numbers M and N such that, for all $n \geq N$, $x_n - y_n \geq \frac{1}{M}$. In other words, for all $n \geq N$, $x_n \geq y_n + \frac{1}{M} > y_n$. Because we have reached a contradiction, we conclude that $x \leq y$. $\square$

There is no analogue to Proposition 9.5 involving strict inequalities; even if x and y are represented by sequences whose terms satisfy $x_n < y_n$ for all n, the best we can hope for is that $x \leq y$. For example, consider the real number x represented by $\{-\frac{1}{n}\}$ and the real number y represented by $\{\frac{1}{n}\}$. Clearly $-\frac{1}{n} < \frac{1}{n}$ for all n, but x is not less than y. The two sequences are in fact equivalent and hence represent the same real number, namely 0.

We can use Proposition 9.5 to establish the Archimedean principle for $\mathbb{R}$. We first stated this principle in Proposition 1.6. We state it again here for convenience.

Proposition 9.6 (Archimedean Principle). *Let $x \in \mathbb{R}$ with $x > 0$. Then there exists $K \in \mathbb{N}$ such that $x > \frac{1}{K}$.*

Proof. Let $\{x_n\}$ be any Cauchy sequence representing x. By the definition of positivity of x, there exist natural numbers M and N such that, for all $n \geq N$, $x_n \geq \frac{1}{M}$. We now apply Proposition 9.5; because $\{x_n\}$ represents x, $\{\frac{1}{M}, \frac{1}{M}, \frac{1}{M} \dots\}$ represents $\frac{1}{M}$, and for all $n \geq N$, $x_n \geq \frac{1}{M}$, the same inequality holds for the real numbers x and $\frac{1}{M}$, that is, $x \geq \frac{1}{M}$. Thus $x > \frac{1}{2M}$, establishing the result with $K = 2M$. □

9.2.5 Sequences of real numbers and completeness

Before we become too snug, we observe that all we have shown so far is that $\mathbb{R}$ satisfies the ordered field axioms, just like $\mathbb{Q}$. Our entire purpose in constructing $\mathbb{R}$ was to obtain an ordered field having the desirable property of completeness not possessed by $\mathbb{Q}$.

At first glance it might seem that there is nothing left to prove. After all, didn't we construct $\mathbb{R}$ by adding to $\mathbb{Q}$ all limits of Cauchy sequences? Not quite. We only added limits of Cauchy sequences of *rational numbers*. We must show that if $\{x_n\}$ is an arbitrary Cauchy sequence of *real numbers*, then there exists $x \in \mathbb{R}$ such that $\lim_{n\to\infty} x_n = x$. There is a second issue; we have not actually proved that if x is a real number and $\{x_n\}$ represents x, then $\lim_{n\to\infty} x_n = x$. The reason we have not yet proved this proposition is because, until the last subsection, we had not defined an order relation on $\mathbb{R}$ and thus expressions like $|x_n - x|$ did not yet have any meaning. Now all the notions in the definition of a limit from Chapter 8 make sense and we are ready to prove this proposition.

Proposition 9.7. *Let $x \in \mathbb{R}$ and suppose $\{x_n\}$ is a Cauchy sequence of rational numbers representing x. Then $\lim_{n\to\infty} x_n = x$.*

Proof. Let ε be a positive real number. We must show that there exists $N \in \mathbb{N}$ such that, for all $n \geq N$, $|x_n - x| < \varepsilon$. By the Archimedean principle, there exists $M \in \mathbb{N}$ such that $\varepsilon > \frac{1}{M}$. Because $\{x_n\}$ is Cauchy, there exists N such that, for all $m, n \geq N$, $|x_n - x_m| < \frac{1}{M}$. Fix n and consider the sequence $\{x_n - x_m\}$. This sequence is Cauchy and represents $x_n - x$. We now apply Proposition 9.5. For all $n \geq N$,

$$-\frac{1}{M} < x_n - x_m < \frac{1}{M}.$$

We can think of this inequality as relating three sequences of rational numbers. Two are constant sequences $\{\frac{1}{M}, \frac{1}{M}, \frac{1}{M}, \dots\}$ and $\{-\frac{1}{M}, -\frac{1}{M}, -\frac{1}{M}, \dots\}$, representing $\frac{1}{M}$ and $-\frac{1}{M}$, respectively. The third is the sequence $\{x_n - x_m\}$ for fixed $n \geq N$, representing $x_n - x$. We may thus conclude that

$$-\frac{1}{M} \leq x_n - x \leq \frac{1}{M}$$

for all $n \geq N$. Because $\varepsilon > \frac{1}{M}$, we find that, for all $n \geq N$, $|x_n - x| < \varepsilon$, as desired. □

This proposition has an important corollary. In simple language, it says that given any real number x, we can find a rational number y as close as we like to x.

Corollary 9.1 (Density of the rationals). *Let $x \in \mathbb{R}$ and let $\varepsilon > 0$ be given. Then there exists $y \in \mathbb{Q}$ such that $|x - y| < \varepsilon$.*

Proof. Let $\{x_n\}$ be any Cauchy sequence of rational numbers representing x. Because $\lim_{n\to\infty} x_n = x$, for the given $\varepsilon > 0$, there exists $N \in \mathbb{N}$ such that, for all $n \geq N$, $|x_n - x| < \varepsilon$. We may therefore take y to be x_N. □

We are ready to prove our main theorem.

Theorem 9.4. *$\mathbb{R}$ is complete.*

Proof. Let $\{x_n\}$ be a Cauchy sequence of real numbers. We must show that there exists a real number y such that $\lim_{n\to\infty} x_n = y$. How can we show the existence of a real number? Proposition 9.7 tells us that any Cauchy sequence of rationals actually converges to a real number. Thus we should use the Cauchy sequence $\{x_n\}$ to obtain a Cauchy sequence $\{y_n\}$ of rational numbers close to the terms of $\{x_n\}$. The Cauchy sequence $\{y_n\}$ determines a real number y, and it actually converges to that y it represents. We then need only show that $\{x_n\}$ converges to the same limit y.

We give all the details. For each $n \in \mathbb{N}_0$, by the density of the rationals (Corollary 9.1), there exists y_n rational such that $|x_n - y_n| < \frac{1}{n+1}$. We claim that the sequence $\{y_n\}$ is Cauchy. Let $\varepsilon > 0$ be given. By the Archimedean principle, there exists $K \in \mathbb{N}$ such that $\frac{1}{K} < \frac{\varepsilon}{3}$. Now, because $\{x_n\}$ is Cauchy, there exists $N_1 \in \mathbb{N}$ such that, for all $m, n \geq N_1$, $|x_n - x_m| < \frac{\varepsilon}{3}$. Let $N = \max\{K, N_1\}$. Then for all $m, n \geq N$,

$$\begin{aligned} |y_n - y_m| &= |(y_n - x_n) + (x_n - x_m) + (x_m - y_m)| \\ &\leq |y_n - x_n| + |x_n - x_m| + |x_m - y_m| \\ &< \frac{1}{n+1} + \frac{\varepsilon}{3} + \frac{1}{m+1} \\ &< \frac{1}{K} + \frac{\varepsilon}{3} + \frac{1}{K} \\ &< \frac{\varepsilon}{3} + \frac{\varepsilon}{3} + \frac{\varepsilon}{3} = \varepsilon. \end{aligned}$$

We now have a Cauchy sequence of rational numbers; hence it represents some real number y and, in fact, has limit y. We claim that $\{x_n\}$ has the same limit y. Let $\varepsilon > 0$ be given, and let K be a natural number with $\frac{1}{K} < \frac{\varepsilon}{2}$. Because $\lim_{n\to\infty} y_n = y$, there exists $N_1 \in \mathbb{N}$ such that, for all $n \geq N_1$, $|y_n - y| < \frac{\varepsilon}{2}$. Let

$N = \max\{N_1, K\}$. Then for all $n \geq N$,

$$\begin{aligned} |x_n - y| &= |(x_n - y_n) + (y_n - y)| \\ &\leq |x_n - y_n| + |y_n - y| \\ &< \frac{1}{n+1} + \frac{\varepsilon}{2} \\ &< \frac{1}{K} + \frac{\varepsilon}{2} \\ &< \frac{\varepsilon}{2} + \frac{\varepsilon}{2} = \varepsilon. \end{aligned}$$

Thus $\{x_n\}$ converges to y and the proof is complete. □

9.3 Problems

1. Consider the sequence s whose n-th term is

$$s_n = \sum_{k=0}^{n} \frac{1}{k!}.$$

   Is s convergent?

2. For each $n \in \mathbb{N}$, let

$$H_n = \sum_{k=1}^{n} \frac{1}{k}.$$

   H_n is called the *n-th harmonic number*. Prove that

$$H_{2^m} \geq 1 + \frac{m}{2}.$$

   Is $\{H_n\}$ convergent?

3. Consider the sequence p defined for $n \geq 2$ by

$$p_n = \prod_{k=2}^{n} \left(1 - \frac{1}{k^2}\right).$$

   Is p convergent?

4. For each of the statements below, either prove the statement if it is true or provide a counterexample if it is false.

   (a) If $\{x_n\}$ is a Cauchy sequence of rational numbers representing x and if $x_n < 0$ for all n, then $x < 0$.

   (b) Suppose $x, y \in \mathbb{R}$ and $x > y$. Then if $\{x_n\}$ is any Cauchy sequence representing x and $\{y_n\}$ is any Cauchy sequence representing y, then $x_n > y_n$ for all n.

5. Suppose $\{a_n\}$ is a sequence of real numbers with $a_n = a_{n-1}^2$ for all $n \geq 1$. Determine how the choice of first term a_0 affects the convergence of $\{a_n\}$.

6. Let $a = \{a_n\}$ be a sequence of real numbers and let $A = \{a_n : n \in \mathbb{N}_0\}$. We define $\sup a_n$ to be the supremum of A, and we define $\inf a_n$ to be the infimum of the set A. For each of the statements below, either give a proof if it is true or give a counterexample if it is false.

   (a) If a converges to L, then $L = \sup a_n$.

   (b) If $\sup a_n = \infty$, a is divergent.

   (c) If $\sup a_n$ and $\inf a_n$ are finite, then a is convergent.

   (d) If a is convergent, $\sup a_n$ and $\inf a_n$ are finite.

   (e) For any two sequences a and b of real numbers, $\sup(a_n + b_n) = \sup a_n + \sup b_n$.

   (f) Let $c \in \mathbb{R}$. Then $\sup(ca_n) = c \sup a_n$.

7. Is $\mathbb{C}$ complete? (You may want to use Problem 10 in Chapter 8.)

Programming Project. Write a program that takes a natural number K as input and returns a decimal approximation of its cube root $\sqrt[3]{K}$. Make your program terminate when the floating point accuracy has been exceeded. *Prove that your algorithm in fact generates a sequence converging to* $\sqrt[3]{K}$.

Chapter 10

Series, Part 1

10.1 Basic Notions

10.1.1 Definitions

In standard English, the words *sequence* and *series* are sometimes used interchangeably. It would be just as correct to describe the *sequence* of steps one must follow to serve a tennis ball as it would be to describe the *series* of steps one must follow. In mathematics, on the other hand, a sequence and a series are not the same thing. Roughly speaking, a sequence is an ordered list whereas a series is the (formal) sum of terms in a sequence. For example,

$$1, \frac{1}{2}, \frac{1}{4}, \frac{1}{8}, \ldots$$

is a sequence, but

$$1 + \frac{1}{2} + \frac{1}{4} + \frac{1}{8} + \ldots$$

is a series. The reason we have said that a series is the *formal* sum of terms in a sequence is that we have not yet said what it means to add up infinitely many things. Sometimes such a sum will make sense. For example, we will see that it makes sense to say that the series above sums to, or *converges* to 2. On the other hand, many such sums do not make sense. For example, we will see that the series

$$1 - 1 + 1 - 1 + 1 - 1 + \ldots$$

doesn't converge. Of course, we need a precise definition.

Definition 10.1. Let $\{a_n\}$ be a sequence of real numbers. The **series** $\sum_{n=0}^{\infty} a_n$ is the formal sum of the terms of the sequence. For each $N \in \mathbb{N}_0$, we define the N-th **partial sum** of the series by $A_N = \sum_{n=0}^{N} a_n$. We say the series $\sum_{n=0}^{\infty} a_n$ **converges** if the sequence $\{A_N\}$ has a limit L, and we write $\sum_{n=0}^{\infty} a_n = L$. Otherwise we say the series **diverges**.

Let's begin our exploration of this definition with the two series above.

Example 10.1. Consider first

$$1+\frac{1}{2}+\frac{1}{4}+\frac{1}{8}+\ldots=\sum_{n=0}^{\infty}\frac{1}{2^n}. \tag{10.1}$$

Does this series converge? Our definition requires us to consider the corresponding sequence $\{A_N\}$ of partial sums. Let's try to understand this sequence. We begin by generating the first few terms.

$$\begin{aligned}
A_0 &= \sum_{n=0}^{0}\frac{1}{2^n}=1\\
A_1 &= \sum_{n=0}^{1}\frac{1}{2^n}=1+\frac{1}{2}=\frac{3}{2}\\
A_2 &= \sum_{n=0}^{2}\frac{1}{2^n}=1+\frac{1}{2}+\frac{1}{4}=\frac{7}{4}\\
A_3 &= \sum_{n=0}^{3}\frac{1}{2^n}=1+\frac{1}{2}+\frac{1}{4}+\frac{1}{8}=\frac{15}{8}.
\end{aligned}$$

We conjecture that

$$A_N=\sum_{n=0}^{N}\frac{1}{2^n}=\frac{2^{N+1}-1}{2^N}=2-\frac{1}{2^N}. \tag{10.2}$$

This conjecture is, in fact, true. We leave the easy proof to the next exercise.

We may now determine whether our series converges. The definition says the series (10.1) converges to the number L if $\lim_{N\to\infty} A_N$ exists and equals L. Because

$$\lim_{N\to\infty} A_N=\lim_{N\to\infty}\left(2-\frac{1}{2^N}\right)=2,$$

we conclude that our original series converges to 2.

Exercise 10.1. Prove (10.2). Several methods are possible; one method is to use induction on N.

Example 10.2. Next we consider the series

$$1-1+1-1+\ldots=\sum_{n=0}^{\infty}(-1)^n. \tag{10.3}$$

We form the corresponding sequence of partial sums:

$$\begin{aligned} A_0 &= \sum_{n=0}^{0}(-1)^n = 1 \\ A_1 &= \sum_{n=0}^{1}(-1)^n = 1-1=0 \\ A_2 &= \sum_{n=0}^{2}(-1)^n = 1-1+1=1 \\ A_3 &= \sum_{n=0}^{3}(-1)^n = 1-1+1-1=0. \end{aligned}$$

We easily see (by an induction proof that we can do in our heads at this point) that

$$A_N = \begin{cases} 1 & \text{if } N \text{ is even} \\ 0 & \text{if } N \text{ is odd.} \end{cases}$$

Because this sequence of partial sums does not have a limit, the series $\sum_{n=0}^{\infty}(-1)^n$ diverges.

Exercise 10.2. Write out the first four terms in the sequence $\{A_N\}$ of partial sums for the series $\sum_{n=0}^{\infty}\left(\frac{1}{n+1}-\frac{1}{n+2}\right)$. Find and prove a general formula for A_N. Does the series converge?

10.1.2 Exploring the sequence of partial sums graphically and numerically

Because the two series considered in the previous subsection were quite simple, we could easily write down a formula for the N-th term of the sequence of partial sums to see if this sequence had a limit. In many cases, however, writing down such a formula is difficult or impossible. We can still get a feeling for the behavior of the sequence of partial sums by exploring it numerically and graphically, and the insights we gain can inform the proof techniques we choose.

Example 10.3. Consider the series

$$\sum_{n=0}^{\infty}\frac{(-1)^n}{(n+1)^2}. \tag{10.4}$$

In order to determine whether the series converges or diverges, we must examine the associated sequence of partial sums. By definition,

$$\begin{aligned}
A_0 &= \sum_{n=0}^{0} \frac{(-1)^n}{(n+1)^2} = 1 \\
A_1 &= \sum_{n=0}^{1} \frac{(-1)^n}{(n+1)^2} = 1 + \frac{(-1)^1}{2^2} = 1 - \frac{1}{4} = \frac{3}{4} \\
A_2 &= \sum_{n=0}^{2} \frac{(-1)^n}{(n+1)^2} = 1 - \frac{1}{4} + \frac{1}{9} = \frac{31}{36}.
\end{aligned}$$

Generating partial sums for this series by hand becomes tedious very quickly. Instead, we write a few lines of Python code to generate and plot the first 50 partial sums.

```
# Define a list initially containing only the first partial sum,
    which is 1
partialSums=[1]

# Each time through the loop, add to the previous partial sum
# the next term in the series
for N in range(1,50):
    partialSums.append(partialSums[N-1]+(-1.0)**(N)/((N+1)**2))

print(partialSums)
plot(partialSums,".")
show()
```

Figure 10.1 shows the plot. The plot certainly suggests that the sequence of partial sums is convergent, but we can see more. It looks as if the partial sums with even index form a decreasing sequence that is bounded below and the partial sums with odd index form an increasing sequence that is bounded above. Furthermore, these sequences appear to have a common limit. From the numerical output, we see that the last three partial sums are

$$0.822254538698166, \quad 0.8226710318260294, \quad 0.8222710318260295,$$

suggesting that the partial sums approach some real number near 0.822.

Of course this numerical and graphical evidence simply leads us to conjecture that the series (10.4) converges. We still require a proof. We postpone the proof until a later section. At this point, our aims are simply to understand clearly the definition of the convergence of a series and to explore examples.

Example 10.4. Next we consider

$$\sum_{n=0}^{\infty} \frac{n}{n+1}.$$

Figure 10.1: First 50 partial sums for Example 10.3.

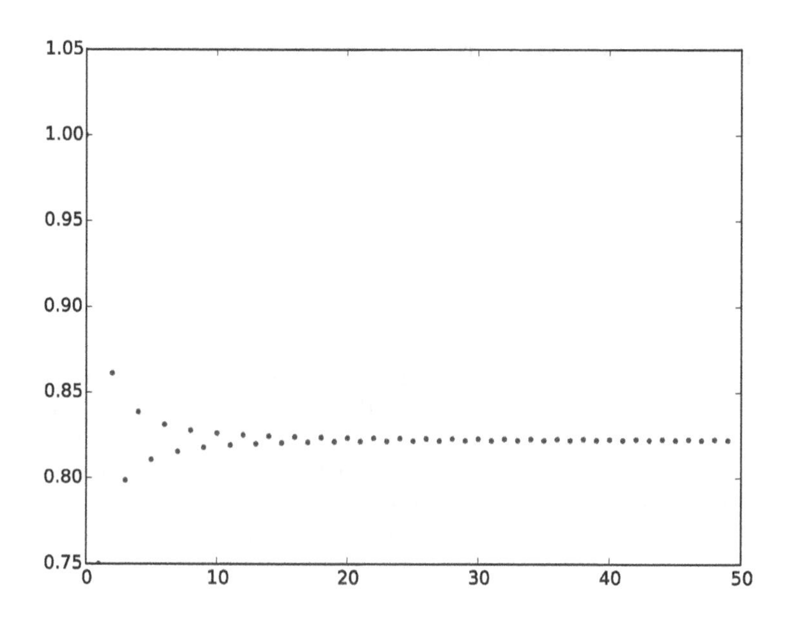

Observe,

$$\begin{aligned} A_1 &= \frac{1}{2} \\ A_2 &= \frac{1}{2} + \frac{2}{3} \\ A_3 &= \frac{1}{2} + \frac{2}{3} + \frac{3}{4} \\ A_4 &= \frac{1}{2} + \frac{2}{3} + \frac{3}{4} + \frac{4}{5}. \end{aligned}$$

Notice that to go from A_N to A_{N+1}, we must add $\frac{N+1}{N+2}$, which is very close to 1 for large N. Thus it is reasonable to conjecture that the sequence $\{A_N\}$ is unbounded, hence divergent.

We prove this claim. Let $K > 0$ be given. We must find N so that $A_N > K$. We can obtain a very simple lower bound for A_N; observe,

$$\begin{aligned} A_2 &= \frac{1}{2} + \frac{2}{3} > \frac{1}{2} + \frac{1}{2} = \frac{2}{2} \\ A_3 &= \frac{1}{2} + \frac{2}{3} + \frac{3}{4} > \frac{1}{2} + \frac{1}{2} + \frac{1}{2} = \frac{3}{2}, \end{aligned}$$

and so on. An easy induction argument shows that, for all $n > 2$, $A_N > \frac{N}{2}$. Because $\frac{N}{2}$ is greater than K if $N > 2K$, our claim follows. Because the sequence $\{A_N\}$ is divergent, the series $\sum_{n=0}^{\infty} \frac{n}{n+1}$ is divergent.

Exercise 10.3. Write the code to generate and plot the first 50 partial sums of the series $\sum_{n=0}^{\infty} \frac{n}{n+1}$.

10.1.3 Basic properties of convergent series

The examples we have considered are of the form $\sum_{n=0}^{\infty} a_n$ for

(a) $a_n = 2^{-n}$,

(b) $a_n = (-1)^n$,

(c) $a_n = (-1)^n/(n+1)^2$,

(d) $a_n = n/(n+1)$.

You may have noticed that in the case of the two divergent series (associated with (b) and (d)), the terms a_n themselves do not have limit zero, whereas in the case of the two convergent series (associated with (a) and (c)), the terms a_n do have limit zero. We will see shortly that the question of convergence of a series can not be so easily settled; there are many divergent series whose terms have limit zero. We have, however, discovered a very important *necessary condition* for convergence.

Proposition 10.1. *Let* $\sum_{n=0}^{\infty} a_n$ *be a series of real numbers. If* $\sum_{n=0}^{\infty} a_n$ *converges, then* $\lim_{n\to\infty} a_n = 0$.

Proof. Let $A_N = \sum_{n=0}^{N} a_n$ be the N-th partial sum of the series. By hypothesis, $\{A_N\}$ is a convergent sequence of real numbers; denote its limit by L. Observe that, for $N \geq 1$,

$$a_N = A_N - A_{N-1}. \tag{10.5}$$

Because $\lim_{N\to\infty} A_N = L$ and $\lim_{N\to\infty} A_{N-1} = L$, the limit of the right-hand side of (10.5) exists. Thus

$$\lim_{N\to\infty} a_N = \lim_{N\to\infty} (A_N - A_{N-1}) = \lim_{N\to\infty} A_N - \lim_{N\to\infty} A_{N-1} = L - L = 0,$$

as claimed. ☐

This simple necessary condition for convergence gives us a very easy way to show that certain series are divergent; if the sequence $\{a_n\}$ does not have limit 0, the series $\sum_{n=0}^{\infty} a_n$ diverges.

Exercise 10.4. Consider again the series $\sum_{n=0}^{\infty} \frac{n}{n+1}$. In light of the above discussion, give a second proof of its divergence.

We close this section with an easy proposition about convergent series whose proof follows from properties of limits.

Proposition 10.2. *Let* $\sum_{n=0}^{\infty} a_n$ *and* $\sum_{n=0}^{\infty} b_n$ *be convergent series of real numbers and let* $c \in \mathbb{R}$.

(a) $\sum_{n=0}^{\infty} ca_n$ *is convergent with sum* $c\sum_{n=0}^{\infty} a_n$.

(b) $\sum_{n=0}^{\infty}(a_n + b_n)$ *is convergent with sum* $\sum_{n=0}^{\infty} a_n + \sum_{n=0}^{\infty} b_n$.

Proof. We prove part (a) and leave the proof of part (b) to the reader.

Let $\{A_N\}$ denote the sequence of partial sums associated with $\sum_{n=0}^{\infty} a_n$ and let $\{S_N\}$ denote the sequence of partial sums associated with $\sum_{n=0}^{\infty} ca_n$. By the distributive law, for every N,

$$S_N = \sum_{n=0}^{N} ca_n = c\sum_{n=0}^{N} a_n = cA_N.$$

By hypothesis, $\sum_{n=0}^{\infty} a_n$ converges; thus there exists a real number L such that $\lim_{N\to\infty} A_N = L$. Thus by part (a) of Proposition 8.3, $\{S_N\} = \{cA_N\}$ converges, with limit cL. In other words, $\sum_{n=0}^{\infty} ca_n$ converges, with $\sum_{n=0}^{\infty} ca_n = c\sum_{n=0}^{\infty} a_n$. □

Exercise 10.5. Prove part (b) of Proposition 10.2.

10.1.4 Series that diverge slowly: The harmonic series

If the condition in Proposition 10.1 (that $\lim_{n\to\infty} a_n = 0$) were both necessary *and sufficient* for the convergence of $\sum_{n=0}^{\infty} a_n$, the study of infinite series would be comparatively simple. Alas, the condition is *not* sufficient. To establish this claim, we must give an example of a series whose terms have limit zero but which is nonetheless divergent. The canonical example is the series whose terms are the reciprocals of the natural numbers:

$$\sum_{n=0}^{\infty} \frac{1}{n+1} = 1 + \frac{1}{2} + \frac{1}{3} + \frac{1}{4} + \dots. \tag{10.6}$$

The series (10.6) is called the *harmonic series.*

Of course, to determine whether the series converges, we must consider the sequence $\{A_N\}$ of partial sums. Observe that, for all N,

$$A_{N+1} = A_N + \frac{1}{N+2} > A_N,$$

so that $\{A_N\}$ is increasing. We know from our study of sequences that an increasing sequence of real numbers converges if and only if it is bounded above. Thus our task is to determine whether or not the sequence $\{A_N\}$ is bounded.

We proceed as above, generating and plotting the first 50 partial sums. See Figure 10.2. Is the sequence of partial sums bounded? It's not clear. Let's look at the first 500 partial sums (Figure 10.3) and the first 5000 partial sums (Figure 10.4). We find that

$$\begin{aligned} A_{49} &= 4.499205338329423 \\ A_{499} &= 6.79282342999052 \\ A_{4999} &= 9.094508852984404. \end{aligned}$$

Figure 10.2: First 50 partial sums of the harmonic series.

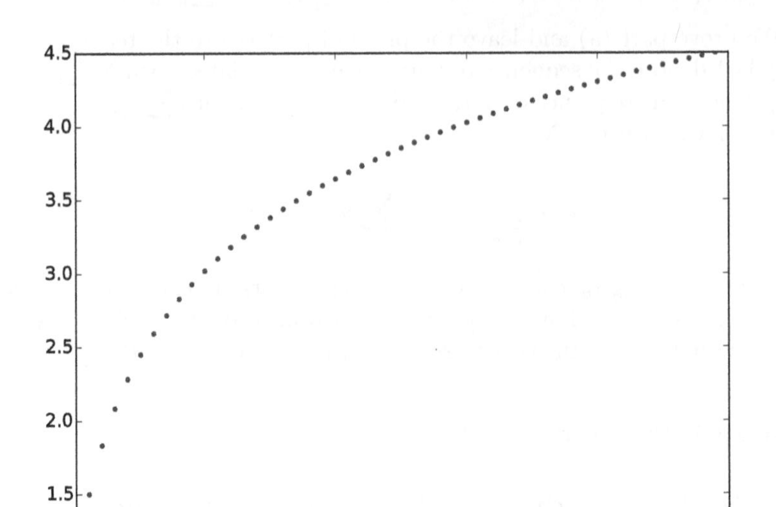

Figure 10.3: First 500 partial sums of the harmonic series.

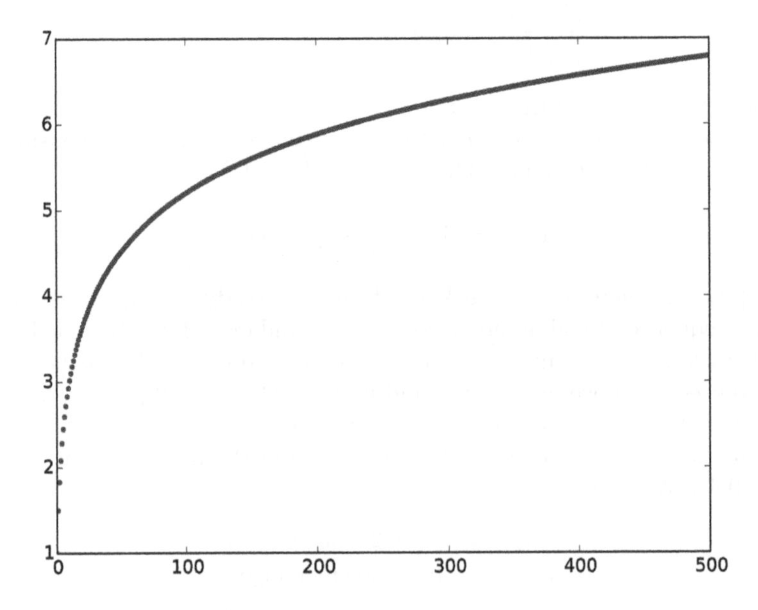

Figure 10.4: First 5000 partial sums of the harmonic series.

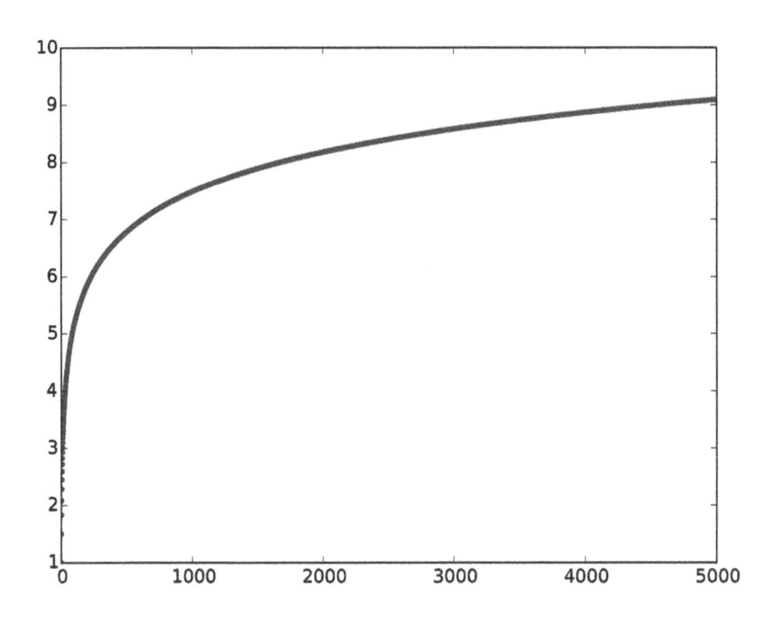

Although at first these data may not seem to shed much light on the situation, they actually contain a clue; the function $N \mapsto A_N$ seems to have the property that increasing the input by a factor of 10 (i.e., considering the fiftieth term, then the five hundredth, then the five thousandth) seems to add a nearly constant amount of about 2.3 to the function. We know from our pre-calculus days that logarithmic functions are defined by this property. We thus suspect that A_N is approximately logarithmic in N. Furthermore, because logarithmic functions are unbounded, we suspect that the same is true of the sequence $\{A_N\}$. If we can prove this conjecture, we will be able to conclude that the harmonic series is in fact divergent.

Exercise 10.6. Write the Python code necessary to generate the plots above and the numerical values of A_{49}, A_{499}, A_{4999}, and A_{49999}.

We now prove that the sequence $\{A_N\}$ is unbounded. Let $K > 0$ be given. We must show that there exists N such that $A_N > K$. We will consider those partial sums with index one less than a power of 2. Each of these involves twice as many terms as its predecessor and ends with a term that is a negative power of 2. Here are the first few.

$$\begin{aligned} A_0 &= 1 = \frac{2}{2} \\ A_1 &= 1 + \frac{1}{2} = \frac{3}{2} \\ A_3 &= 1 + \frac{1}{2} + \left(\frac{1}{3} + \frac{1}{4}\right) \geq \frac{3}{2} + \frac{1}{2} = \frac{4}{2} \\ A_7 &= A_3 + \left(\frac{1}{5} + \frac{1}{6} + \frac{1}{7} + \frac{1}{8}\right) \geq \frac{4}{2} + \frac{4}{8} = \frac{5}{2}. \end{aligned}$$

A simple induction argument shows that

$$A_{2^k-1} \geq \frac{k+2}{2}.$$

Note that

$$\frac{k+2}{2} > K \iff k > 2K - 2.$$

Set $k_0 = 2\lceil K \rceil - 1$. Then if $N = 2^{k_0} - 1$, $A_N = A_{2^{n_0}-1} > K$. We have shown that the sequence of partial sums associated with the harmonic series is divergent, and hence we have shown that the harmonic series is divergent.

Exercise 10.7. Let A_N denote the N-th partial sum of the harmonic series, as above. How large must N be in order to have $A_N > 10$? $A_N > 100$?

10.2 Infinite Geometric Series

We recall the terminology for a type of sequence considered in Chapter 2.

Definition 10.2. A sequence g is **geometric** if $g_n = ar^n$ for $a, r \in \mathbb{R}$ and $n \in \mathbb{N}_0$. We call r the **common ratio**. An **infinite geometric series** is thus the series $\sum_{n=0}^{\infty} ar^n$.

For which, if any, a and r is an infinite geometric series convergent? We consider the sequence of partial sums

$$S_N = \sum_{n=0}^{N} ar^n.$$

In Problem 6 of Chapter 2 we sought a simple formula for this finite geometric series. The result for $r \neq 1$ is

$$S_N = \frac{a(1 - r^{N+1})}{1 - r}. \tag{10.7}$$

The sequence S_N has a limit if and only if the sequence $\{r^{N+1}\}$ has a limit. It is easy to show that $\{r^{N+1}\}$ has limit 0 if $|r| < 1$ and diverges if $r > 1$ or $r \leq -1$. Thus S_N converges if and only if $|r| < 1$, in which case its limit is $\frac{a}{1-r}$. Observe that if $r = 1$, $S_N = a(N+1)$, which diverges unless $a = 0$. We have therefore proved the following proposition.

Proposition 10.3. *Consider the infinite geometric series $\sum_{n=0}^{\infty} ar^n$ for $a \neq 0$. Then*

(a) if $|r| < 1$, the series converges to $\frac{a}{1-r}$.

(b) if $|r| \geq 1$, the series diverges.

Exercise 10.8. Each of the following series is geometric. Determine whether each is convergent or divergent. Find the sum of each convergent series.

(a) $\displaystyle\sum_{n=0}^{\infty} 3^{-n}$.

(b) $\displaystyle\sum_{n=1}^{\infty} \frac{(-2)^n}{5^{n+1}}$.

(c) $\displaystyle\sum_{n=1}^{\infty} \frac{4^{n-1}}{3^n}$.

(d) $\dfrac{3}{10} + \dfrac{3}{100} + \dfrac{3}{1000} + \dfrac{3}{10000} + \ldots$. How do you express this number as a decimal?

(e) $\displaystyle\sum_{n=0}^{\infty} x^{2n}$ for $|x| < 1$.

10.3 Tests for Convergence of Series

A word on notation: In the propositions that follow, when we are considering a general infinite series, we will often just write $\sum a_n$ instead of the more cumbersome $\sum_{n=0}^{\infty} a_n$. When we consider any specific example, however, we will always include the limits of summation.

At this point, we have a good understanding of the definition of convergence for a series and a complete understanding of infinite geometric series. Our goal now is to develop theorems that can help us settle the question of convergence for much more general series. The easiest series to consider are those whose terms are all of one sign because then the sequence of partial sums is monotone. In this case, by the Monotone Sequence Theorem, we can prove convergence by proving boundedness. The latter is much easier to do for a number of reasons, not the least of which is that it is often nearly impossible to guess the limit of the partial sums. Because of the relative ease of analyzing series of non-negative terms, we often begin our analysis of a general series $\sum a_n$ by considering its sister series $\sum |a_n|$ with non-negative terms. Such an approach is legitimized by the next proposition, which says that if the sister series converges, then so does the original series.

Proposition 10.4. *Let $\{a_n\}$ be a sequence of real numbers. If $\sum |a_n|$ converges, then $\sum a_n$ converges.*

Proof. Let $A_N = \sum_{n=0}^{N} a_n$. We must show that $\{A_N\}$ converges. We will prove this assertion by showing that $\{A_N\}$ is Cauchy.

Let $\varepsilon > 0$ be given. We must show that there exists K such that, for all $N, M \geq K$, $|A_N - A_M| < \varepsilon$. Assume without loss of generality that $N > M$. Let $S_N = \sum_{n=0}^{N} |a_n|$. Observe,

$$\begin{aligned} |A_N - A_M| &= \left| \sum_{n=0}^{N} a_n - \sum_{n=0}^{M} a_n \right| \\ &= \left| \sum_{n=M+1}^{N} a_n \right| \\ &\leq \sum_{n=M+1}^{N} |a_n| \quad \text{(by the triangle inequality)} \\ &= S_N - S_M \\ &= |S_N - S_M|. \end{aligned}$$

Because $\sum |a_n|$ converges, the sequence $\{S_N\}$ is Cauchy. Thus for ε as above, there exists K such that, for all $N, M \geq K$, $|S_N - S_M| < \varepsilon$. Thus for this same K, if $N > M \geq K$, $|A_N - A_M| < \varepsilon$. □

Because we use Proposition 10.4 so often, we develop some relevant terminology.

Definition 10.3. Let $\sum a_n$ be a series of real numbers. If the associate series $\sum |a_n|$ converges, we say that $\sum a_n$ **converges absolutely.**

We often use this terminology to restate Proposition 10.4 as follows: If a series converges absolutely, then it converges. It is incredibly important to note that *the converse of Proposition 10.4 is false.* There are convergent series $\sum a_n$ for which the associate sister series $\sum |a_n|$ diverges. The canonical example is the alternating harmonic series $\sum_{n=1}^{\infty} \frac{(-1)^n}{n}$. We postpone the proof of its convergence until Chapter 14. We already know that the absolute series is divergent, for it is the harmonic series. Such *conditionally convergent* series have some interesting and very counterintuitive properties that will be discussed in Chapter 14. In this chapter we mostly treat the more mundane absolutely convergent series.

The next proposition on series of non-negative terms is incredibly useful because it allows us to prove the convergence or divergence of a series by relating its terms to those of another series whose convergence or divergence is known.

Proposition 10.5 (Comparison Test). *Let $\{a_n\}$ and $\{b_n\}$ be sequences of non-negative real numbers. Suppose $0 \leq a_n \leq b_n$.*

(a) If $\sum b_n$ converges, then $\sum a_n$ converges.

(b) If $\sum a_n$ diverges, then $\sum b_n$ diverges.

Proof. The proof of (a) is almost identical to the proof of Proposition 10.4 and is left to the reader as an exercise.

Consider (b). Let $A_N = \sum_{n=0}^{N} a_n$ and let $B_N = \sum_{n=0}^{N} b_n$. Because the terms a_n are non-negative, the sequence $\{A_N\}$ is non-decreasing. If it were bounded above, it would be convergent by the Monotone Sequence Theorem, and the series $\sum a_n$ would be convergent. Because the series is not convergent, we may conclude that the sequence $\{A_N\}$ is unbounded. We claim that $\{B_N\}$ is also unbounded. Indeed, let $M \in \mathbb{R}$ be given. Because $\{A_N\}$ is unbounded, there exists N such that $A_N > M$. But then

$$\begin{aligned} B_N &= \sum_{n=0}^{N} b_n \\ &\geq \sum_{n=0}^{N} a_n \\ &= A_N \\ &> M. \end{aligned}$$

Because the sequence $\{B_N\}$ is unbounded, the series $\sum b_n$ is divergent. □

Exercise 10.9. Prove part (a) of Proposition 10.5.

Exercise 10.10. Prove that the conclusions of Proposition 10.5 still hold if we replace the hypothesis $a_n \leq b_n$ with the hypothesis that there exists n_0 such that, for all $n \geq n_0$, $a_n \leq b_n$.

Exercise 10.11. Consider $\sum_{n=1}^{\infty} \frac{1}{n^n}$. Prove that this series is convergent by comparing it to a convergent geometric series.

In order to use the comparison test, we need a larger class of series whose convergence or divergence we can establish directly. We therefore consider the p-series.

Theorem 10.1 (p-Series Test). *Let $p > 0$ and consider the* **p-series** $\sum_{n=1}^{\infty} \frac{1}{n^p}$ *(or, with zero-indexing, $\sum_{n=0}^{\infty} \frac{1}{(n+1)^p}$). This series converges if $p > 1$ and diverges if $0 < p \leq 1$.*

This proof will resemble our proof that the harmonic series is divergent, in which we obtained a lower bound for partial sums involving a number of terms that is a power of 2. Thus we need a lemma estimating such partial sums for p-series.

Lemma 10.1. *Let $A_N = \sum_{n=1}^{N} \frac{1}{n^p}$. Then*

$$1 + \frac{1}{2}\sum_{n=1}^{N}(2^{1-p})^n \leq A_{2^N} \tag{10.8}$$

and

$$A_{2^N-1} \leq 1 + \sum_{n=1}^{N-1}(2^{1-p})^n. \tag{10.9}$$

Proof. Each inequality is proved by induction on N.

Consider first (10.8). When $N = 1$,

$$\begin{aligned} A_{2^1} &= 1 + \frac{1}{2^p} \\ &= 1 + \frac{1}{2} \cdot 2^{1-p} \end{aligned}$$

and the result holds.

Now suppose the result holds for $N \geq 1$ and consider $N + 1$.

$$\begin{aligned} A_{2^{N+1}} &= \sum_{n=1}^{2^N} \frac{1}{n^p} + \sum_{n=2^N+1}^{2^{N+1}} \frac{1}{n^p} \\ &\geq 1 + \frac{1}{2}\sum_{n=1}^{N}(2^{1-p})^n + \sum_{n=2^N+1}^{2^{N+1}} \frac{1}{(2^{N+1})^p} \\ &= 1 + \frac{1}{2}\sum_{n=1}^{N}(2^{1-p})^n + \frac{1}{2} \cdot \frac{2^{N+1}}{(2^{N+1})^p} \\ &= 1 + \frac{1}{2}\sum_{n=1}^{N}(2^{1-p})^n + \frac{1}{2}(2^{1-p})^{N+1} \\ &= 1 + \frac{1}{2}\sum_{n=1}^{N+1}(2^{1-p})^n. \end{aligned}$$

Thus the inequality holds for $N+1$ and hence, by the principle of mathematical induction, for all N.

The proof of the second inequality is also by induction. When $N = 1$,

$$\begin{aligned} A_{2^1-1} &= A_1 \\ &= 1 \\ &= 1 + \sum_{n=1}^{0}(2^{1-p})^n, \end{aligned}$$

because the sum is empty and equals 0. Thus the result holds for $N = 1$.

Suppose now that the result holds for $N \geq 1$ and consider $N+1$. Then

$$\begin{aligned} A_{2^{N+1}-1} &= \sum_{n=1}^{2^N-1} \frac{1}{n^p} + \sum_{n=2^N}^{2^{N+1}-1} \frac{1}{n^p} \\ &\leq 1 + \sum_{n=1}^{N-1} (2^{1-p})^n + \sum_{n=2^N}^{2^{N+1}-1} \frac{1}{(2^N)^p} \\ &= 1 + \sum_{n=1}^{N-1} (2^{1-p})^n + \frac{2^N}{(2^N)^p} \\ &= 1 + \sum_{n=1}^{N-1} (2^{1-p})^n + (2^{1-p})^N \\ &= 1 + \sum_{n=1}^{N} (2^{1-p})^n, \end{aligned}$$

and the result holds for $N+1$. By the principle of mathematical induction, it holds for all N. □

We may now prove our theorem on p-series.

Proof. We prove first that the series $\sum_{n=1}^{\infty} \frac{1}{n^p}$ converges for $p > 1$. Let $A_N = \sum_{n=1}^{N} \frac{1}{n^p}$. Clearly $\{A_N\}$ is non-decreasing, so it suffices to show that $\{A_N\}$ is bounded. Indeed, for any N, $N \leq 2^N - 1$, and so by the previous lemma,

$$A_N \leq A_{2^N-1} \leq 1 + \sum_{n=1}^{N-1} (2^{1-p})^n.$$

The sum on the right is a partial sum of a geometric series with $r = 2^{1-p}$ and so is convergent if $|r| < 1$. Because $p > 1$, $2^1 < 2^p$, or, equivalently, $2^{1-p} < 1$. We have therefore shown that, for all N,

$$A_N \leq 1 + \sum_{n=1}^{\infty} (2^{1-p})^n = 1 + \frac{2^{1-p}}{1 - 2^{1-p}}.$$

Thus by the Monotone Sequence Theorem, $\sum_{n=1}^{\infty} \frac{1}{n^p}$ is convergent.

We can approach the proof of the divergence of $\sum_{n=1}^{\infty} \frac{1}{n^p}$ for $0 < p \leq 1$ in several ways. One approach is to prove that A_N is unbounded by using the lower bound for A_{2^N} obtained in the previous lemma. A second approach (which is the one we use here) simply uses the comparison test and the result we have already proved that the p-series for $p = 1$ is divergent. Indeed, note that, for $p < 1$, for $n \geq 1$, $n^{1-p} \geq 1$, or, equivalently,

$$\frac{1}{n^p} \geq \frac{1}{n}.$$

Thus because $\sum_{n=1}^{\infty} \frac{1}{n}$ is divergent, by the comparison test, $\sum_{n=1}^{\infty} \frac{1}{n^p}$ is divergent if $p < 1$. □

The p-series test and the comparison test together allow us to determine the convergence or divergence of any series whose n-th term is a rational function of n. The next exercise asks you to consider three concrete examples; the problems at the end of the chapter ask you to formulate and prove a general statement.

Exercise 10.12. For each of the series below, use the comparison test to prove the convergence or divergence of the given series.

(a) $\displaystyle\sum_{n=1}^{\infty} \frac{n}{n^3+1}$.

(b) $\displaystyle\sum_{n=2}^{\infty} \frac{n}{n^2-1}$.

(c) $\displaystyle\sum_{n=1}^{\infty} \frac{n}{\sqrt{n^3+1}}$.

Exercise 10.13. Give a proof that $\sum_{n=1}^{\infty} \frac{1}{n^p}$ diverges for $p < 1$ using the lower bound (10.8) for A_{2^N}.

Of course, we can compare a series with non-negative terms to a p-series even if its n-th term is not a rational function of n; all we need do is establish an inequality.

Example 10.5. Consider $\sum_{n=0}^{\infty} \frac{1}{n!}$. Observe that for all $n \geq 2$,

$$n! \geq n(n-1) = n^2 - n \geq n^2 - \frac{1}{2}n^2 = \frac{1}{2}n^2.$$

Thus for all $n \geq 2$,

$$\frac{1}{n!} \leq \frac{2}{n^2}.$$

The series $\sum_{n=1}^{\infty} \frac{2}{n^2}$ is convergent because it is a constant multiple of the convergent p-series with $p = 2$. Thus by the comparison test (with the strengthening given in Exercise 10.10), $\sum_{n=0}^{\infty} \frac{1}{n!}$ is convergent.

Several very useful tests for convergence come about by comparing a given series to a geometric series. You may recognize these tests from calculus.

Theorem 10.2 (Ratio Test). *Let $\{a_n\}$ be a sequence of non-zero real numbers. Suppose $\lim_{n\to\infty} \left|\frac{a_n}{a_{n-1}}\right|$ exists and equals L. Then*

(a) If $L < 1$, $\sum a_n$ converges absolutely.

(b) If $L > 1$, $\sum a_n$ diverges.

Note that this theorem says nothing about what happens when $L = 1$. We will see that there are examples of both convergent and divergent series for which $L = 1$.

Proof. Suppose first that $\lim_{n\to\infty}\left|\frac{a_n}{a_{n-1}}\right| = L < 1$. Let ε be a strictly positive real number for which $L < L + \varepsilon < 1$. (Such an ε exists; take, for example, $\varepsilon = (1 - L)/2$.) Because $\lim_{n\to\infty}\left|\frac{a_n}{a_{n-1}}\right| = L$, for this ε, there exists N such that, for all $n \geq N$,

$$\left|\left|\frac{a_n}{a_{n-1}}\right| - L\right| < \varepsilon.$$

This inequality implies that, for all $n > N$,

$$\left|\frac{a_n}{a_{n-1}}\right| < L + \varepsilon \iff |a_n| < |a_{n-1}|(L + \varepsilon). \tag{10.10}$$

Iterating this inequality yields that, for all $n > N$,

$$|a_n| < |a_N|(L + \varepsilon)^{n-N} = \frac{|a_N|}{(L + \varepsilon)^N} \cdot (L + \varepsilon)^n. \tag{10.11}$$

The expression on the far right is the n-th term of a geometric series with common ratio $r = L + \varepsilon$. Because $|r| < 1$, this geometric series is convergent. Thus by the comparison test (and Exercise 10.10), $\sum |a_n|$ is convergent. Then by Proposition 10.4, $\sum a_n$ is convergent as well. We have thus established (a).

Suppose next that $\lim_{n\to\infty}\left|\frac{a_n}{a_{n-1}}\right| = L > 1$ and let $\varepsilon > 0$ be such that $L - \varepsilon > 1$. Because $\lim_{n\to\infty}\left|\frac{a_n}{a_{n-1}}\right| = L$, for this ε, there exists N such that, for all $n \geq N$,

$$\left|\left|\frac{a_n}{a_{n-1}}\right| - L\right| < \varepsilon.$$

This inequality implies that, for all $n > N$,

$$L - \varepsilon < \left|\frac{a_n}{a_{n-1}}\right| \iff |a_{n-1}|(L - \varepsilon) < |a_n|.$$

Iterating as above gives, for all $n > N$,

$$|a_n| > \frac{|a_N|}{(L - \varepsilon)^N} \cdot (L - \varepsilon)^n.$$

Because $L - \varepsilon > 1$, the terms a_n do not have limit 0 (in fact the sequence $\{a_n\}$ is unbounded). Thus by Proposition 10.1, $\sum a_n$ is divergent. □

Exercise 10.14. Give another proof of the convergence of $\sum_{n=0}^{\infty} \frac{1}{n!}$, this time using the ratio test.

The ratio test is a favorite test for convergence because it is so easy to apply. Unfortunately it gives no information when $L = 1$, as the following example illustrates.

Example 10.6. Let $a_n = \frac{1}{n}$ and let $b_n = \frac{1}{n^2}$. Then

$$\begin{aligned}
\lim_{n\to\infty}\left|\frac{a_n}{a_{n-1}}\right| &= \lim_{n\to\infty}\frac{\frac{1}{n}}{\frac{1}{n-1}} \\
&= \lim_{n\to\infty}\frac{n-1}{n} \\
&= \lim_{n\to\infty}\frac{1-\frac{1}{n}}{1} \\
&= 1.
\end{aligned}$$

Similarly, $\lim_{n\to\infty}\left|\frac{b_n}{b_{n-1}}\right| = 1$. Because $\sum a_n$ is divergent and $\sum b_n$ is convergent, we see that the ratio test is inconclusive when $L = 1$.

In fact, the ratio test is always inconclusive when the n-th term is a rational function of n.

Exercise 10.15. Let p and q be polynomials that are not the zero polynomial. Consider $\sum_{n=n_0}^{\infty}\frac{p(n)}{q(n)}$ where n_0 is large enough that neither $p(n)$ nor $q(n)$ is zero for $n \geq n_0$. Show that the ratio test in inconclusive (i.e., that $L = 1$).

The next theorem has a similar flavor to the ratio test and can be proved with a very similar argument that we leave as an exercise.

Theorem 10.3 (Root Test). *Let $\{a_n\}$ be a sequence of real numbers. Suppose $\lim_{n\to\infty}\sqrt[n]{|a_n|}$ exists and equals L. Then*

(a) If $L < 1$, $\sum a_n$ converges absolutely.

(b) If $L > 1$, $\sum a_n$ diverges.

Example 10.7. As an easy example, we obtain a second proof of the convergence of $\sum_{n=1}^{\infty}\frac{1}{n^n}$. We need only observe that

$$\lim_{n\to\infty}\sqrt[n]{\frac{1}{n^n}} = \lim_{n\to\infty}\frac{1}{n} = 0.$$

Because this limit is less than 1, by the root test, $\sum_{n=1}^{\infty}\frac{1}{n^n}$ converges.

Exercise 10.16. Prove Theorem 10.3.

Exercise 10.17. Show that the root test is inconclusive when $L = 1$. You may have to use some techniques you learned in calculus to evaluate some indeterminate limits.

10.4 Representations of Real Numbers

Infinite series may seem esoteric, but you already have considerable informal familiarity with them; every time you describe a real number by giving its decimal representation, you are specifying a convergent series. In this section we explore decimal representations more closely, examine decimal representations of rational numbers, and discuss how to represent real numbers with respect to bases other than 10.

10.4.1 Base 10 representation

Let $0 \leq x < 1$. We consider it a familiar fact that we can write

$$x = 0.a_1a_2a_3a_4\ldots$$

where each a_n is one of the digits 0 through 9. What does such an expression really mean? It is a shorthand notation for the infinite series

$$\frac{a_1}{10} + \frac{a_2}{100} + \frac{a_3}{1000} + \frac{a_4}{10000} + \ldots = \sum_{n=1}^{\infty} a_n 10^{-n}. \tag{10.12}$$

Two obvious questions present themselves:

1. Does every series of the form (10.12) converge to a real number x satisfying $0 \leq x < 1$?

2. Can every real number x satisfying $0 \leq x < 1$ be written in the form (10.12)?

The first proposition answers the first question essentially in the affirmative. (It is possible for the series (10.12) to converge to 1.)

Proposition 10.6. *Consider the infinite series $\sum_{n=1}^{\infty} a_n 10^{-n}$ where each a_n is an integer satisfying $0 \leq a_n \leq 9$. This series converges to a real number x satisfying $0 \leq x \leq 1$.*

Proof. Let $x_N = \sum_{n=1}^{N} a_n 10^{-n}$. Because $\{x_N\}$ is non-decreasing, it is convergent if it is bounded. Observe,

$$\begin{aligned} x_N &= \sum_{n=1}^{N} a_n 10^{-n} \\ &\leq \sum_{n=1}^{N} 9 \cdot 10^{-n} \\ &= 9 \sum_{n=1}^{N} 10^{-n} \\ &\leq 9 \sum_{n=1}^{\infty} 10^{-n} \\ &= 9 \cdot \frac{\frac{1}{10}}{1 - \frac{1}{10}} = 1. \end{aligned}$$

In the above string of inequalities, we have used the fact that $\sum_{n=1}^{N} 10^{-n}$ is a partial sum for a convergent geometric series with positive terms, hence is bounded above by the corresponding infinite geometric series.

Because $\{x_N\}$ is bounded and non-decreasing, there exists $x \in \mathbb{R}$ such that $\lim_{N\to\infty} x_N = x$. Because $0 \leq x_N \leq 1$ and non-strict inequalities are preserved when taking limits, $0 \leq x \leq 1$. □

In order to answer the second question above, we need to describe how to find the a_n in (10.12) given $0 \leq x < 1$. For each such x and each $N \in \mathbb{N}$, let y_N be the largest element of $\mathbb{N}_0$ for which $y_N \leq 10^N x$. Then set x_N equal to $y_N \cdot 10^{-N}$. The numbers x_N give us N-digit decimal approximations to x. Let us clarify with an example.

Example 10.8. Consider $x = \frac{3}{11}$. We find y_N and x_N for $N = 1, 2, 3, 4$.

Consider first $N = 1$. Because $2 \leq 10 \cdot \frac{3}{11}$ but $3 > 10 \cdot \frac{3}{11}$, we obtain $y_1 = 2$ and $x_1 = \frac{y_1}{10} = \frac{2}{10} = 0.2$.

Next, consider $N = 2$. Because $27 \leq 100 \cdot \frac{3}{11}$ but $28 > 100 \cdot \frac{3}{11}$, we obtain $y_2 = 27$ and $x_2 = \frac{y_2}{100} = \frac{27}{100} = 0.27$.

Next, consider $N = 3$. Because $272 \leq 1000 \cdot \frac{3}{11}$ but $273 > 1000 \cdot \frac{3}{11}$, we obtain $y_3 = 272$ and $x_3 = \frac{y_3}{1000} = \frac{272}{1000} = 0.272$.

Finally, consider $N = 4$. Because $2727 \leq 10000 \cdot \frac{3}{11}$ but $2728 > 10000 \cdot \frac{3}{11}$, we obtain $y_4 = 2727$ and $x_4 = \frac{y_4}{1000} = \frac{2727}{10000} = 0.2727$.

How do we recover the digits a_n from the x_n? We set $a_n = 10^n(x_n - x_{n-1})$, where we take $y_0 = x_0 = 0$. You should check that, in the above example, this formula indeed yields $a_1 = 2$, $a_2 = 7$, $a_3 = 2$, and $a_4 = 7$. Observe that, in general,

$$\begin{aligned} a_n &= 10^n(x_n - x_{n-1}) \\ &= 10^n\left(\frac{y_n}{10^n} - \frac{y_{n-1}}{10^{n-1}}\right) \\ &= y_n - 10y_{n-1}. \end{aligned}$$

Because y_n and y_{n-1} are integers, a_n is an integer. Also, because $y_{n-1} \leq 10^{n-1}x$, $10y_{n-1} \leq 10^n x$. Because y_n is the largest element of $\mathbb{N}_0$ with this property, $y_n \geq 10y_{n-1}$, so $a_n \geq 0$. Finally, because $y_{n-1} + 1 > 10^{n-1}x$,

$$a_n < 10^n x - 10(10^{n-1}x - 1) = 10.$$

Thus a_n is indeed one of the digits 0 through 9. Because

$$x_N = \sum_{n=1}^{N}(x_n - x_{n-1}) = \sum_{n=1}^{N} \frac{a_n}{10^n},$$

the a_n really are the digits in the decimal expansion of x_N.

To complete the proof that $\sum_{n=1}^{\infty} \frac{a_n}{10^n} = x$, we show that $\lim_{N\to\infty} x_N = x$. We already know that $\{x_N\}$ is non-decreasing because $x_N - x_{N-1} \geq 0$. We also know that $x_N \leq x$ for all x. Thus $\{x_N\}$ converges to some L. Because

$$y_N \leq 10^N x < y_N + 1,$$

we have

$$x_N \leq x \leq x_N + 10^{-N}.$$

Because non-strict inequalities are preserved when taking limits,

$$L \leq x \leq L$$

and, indeed, $L = x$. We have thus not only proved that every real number $0 \leq x < 1$ has a decimal representation, we have also shown how to obtain one.

Exercise 10.18. Use the procedure outlines above to find the first four digits in the decimal expansion of $x = \frac{4}{7}$.

10.4.2 Base 10 representations of rational numbers

You probably consider it a familiar fact that a number is rational if and only if its decimal representation terminates or repeats. In this subsection, we aim to prove part of this statement. In particular, we will show that, if x has a decimal expansion that infinitely repeats a string of digits, then x is rational. We work up to this goal by first considering some concrete examples.

Example 10.9. Consider the real number x with decimal representation $0.\overline{17} = 0.171717\ldots$. We claim that x is rational.

By definition,

$$x = \frac{1}{10} + \frac{7}{10^2} + \frac{1}{10^3} + \frac{7}{10^4} + \ldots.$$

We would like to group the terms in pairs and write

$$x = \frac{17}{10^2} + \frac{17}{10^4} + \frac{17}{10^6} + \ldots = \sum_{n=1}^{\infty} 17\left(\frac{1}{10^2}\right)^n. \tag{10.13}$$

Is this grouping legitimate? It turns out that for arbitrary series, grouping the terms may change the convergence properties of the series or the value to which it converges. For series of positive terms, however, such grouping is legitimate. See the exercises following this example and the section on groupings and rearrangements in Chapter 14 for a more extensive discussion of these issues.

Taking equation (10.13) as valid, we see that x equals a convergent geometric series with $r = \frac{1}{100}$ and $a = \frac{17}{100}$, and so

$$x = \frac{\frac{17}{100}}{1 - \frac{1}{100}} = \frac{17}{99}.$$

Exercise 10.19. The purpose of this exercise is to illustrate that, if the terms of a series are not all positive, grouping of terms may not be legitimate. Let $a_n = (-1)^n$. Determine whether each of the following series converges or diverges.

(a) $\displaystyle\sum_{n=0}^{\infty} a_n$.

(b) $\displaystyle\sum_{n=0}^{\infty}(a_{2n} + a_{2n+1}) = (a_0 + a_1) + (a_2 + a_3) + (a_4 + a_5) + \ldots$.

(c) $\displaystyle a_0 + \sum_{n=1}^{\infty}(a_{2n-1} + a_{2n}) = a_0 + (a_1 + a_2) + (a_3 + a_4) + (a_5 + a_6) + \ldots$.

Exercise 10.20. Let $\{a_n\}$ be a sequence of *non-negative* real numbers. Consider $b_n = a_{2n}+a_{2n+1}$. Prove that $\sum a_n$ converges if and only if $\sum b_n$ converges.

Exercise 10.21. In each case, find the rational number with the given decimal representation.

(a) $0.\overline{123}$.

(b) $0.123\overline{9}$.

With these examples and exercises behind us, we prove a general result.

Proposition 10.7. *Suppose x is a real number with decimal expansion*

$$0.a_1 \dots a_N \overline{a_{N+1} \dots a_{N+p}}.$$

Then x is rational.

Proof. The proof uses the same idea as the example above; we express x as an infinite series and then observe that the series involves a convergent geometric series. Observe,

$$\begin{aligned} x &= \frac{a_1 \dots a_N}{10^N} + \frac{a_{N+1} \dots a_{N+p}}{10^{N+p}} + \frac{a_{N+1} \dots a_{N+p}}{10^{N+2p}} + \dots \\ &= \frac{a_1 \dots a_N}{10^N} + \sum_{n=1}^{\infty} \frac{a_{N+1} \dots a_{N+p}}{10^{N+np}}. \end{aligned}$$

The last infinite series is a geometric series with $a = \frac{a_{N+1} \dots a_{N+p}}{10^{N+p}}$ and $r = 10^{-p}$. It thus converges to $a/(1-r)$ which is rational because a and r are rational. Because x is the sum of this number and the rational number $\frac{a_1 \dots a_N}{10^N}$, x is itself rational. □

In Chapter 7, we showed that the set of all infinite sequences of zeros and ones is uncountable. By a very similar argument, the set of all sequences $\{a_n\}$ in which each a_n is one of the digits 0 through 9 is uncountable. We pointed out in Chapter 7 that these observations could be the basis for a proof that $\mathbb{R}$ is uncountable, but that we would have to have a deeper understanding of representations of real numbers. We are now at the point where we can complete such an argument. The next exercise leads you through the details.

Exercise 10.22. The purpose of this exercise is to establish that $\mathbb{R}$ is uncountable.

(a) Every element of $[0, 1)$ has a decimal representation $0.a_1a_2a_3\dots$ where a_j is one of the digits 0 through 9. Unfortunately, this fact does not give us a one-to-one correspondence between elements of $[0, 1)$ and sequences $\{a_1, a_2, a_3, \dots\}$ because some real numbers have more than one decimal representation. For example, $0.5 = 0.4\overline{9}$. Describe all elements of $[0, 1)$ having more than one decimal representation.

(b) In light of the above, describe a bijection between $[0, 1)$ and some subset of sequences $\{a_j\}$ in which each a_j is one of the digits 0 through 9.

(c) Prove that $[0, 1)$ is uncountable.

Exercise 10.23. Prove that the set of irrational numbers is uncountable. Thus, in terms of cardinality, there are many more irrational numbers than rational numbers.

10.4.3 Representations in other bases

We saw in Chapter 6 that it can be convenient to represent natural numbers using bases other than 10. We also saw that, given a natural number $b \geq 2$, every natural number has a unique base b representation. In this section we discuss representations of real numbers in bases other than 10. The good news is that base b representations of real numbers work just like decimal (base 10) representations of real numbers. We can prove results analogous to those obtained there using basically the same proof technique. We will omit most of the details and will instead give the essential definitions and a few examples.

Definition 10.4. Let $b \geq 2$ be a number and let x be a real number with $0 \leq x < 1$. A **base b representation** of x is a way of writing $x = \sum_{n=1}^{\infty} a_n b^{-n}$, where each a_n is an integer with $0 \leq a_n \leq b - 1$. By analogy with decimal representations, we often use $0.a_1a_2a_3\ldots_b$ as a short-hand notation for the series.

Example 10.10. We will describe the rational number x with base 3 representation $0.\overline{12}_3$. By definition, this x is the value of the series

$$\frac{1}{3} + \frac{2}{3^2} + \frac{1}{3^3} + \frac{2}{3^4} + \frac{1}{3^5} + \frac{2}{3^6} + \ldots.$$

The discussion of grouping that we had in the previous section applies here as well, so that the series under consideration can be rewritten:

$$\begin{aligned} x &= \left(\frac{1}{3} + \frac{2}{3^2}\right) + \left(\frac{1}{3^3} + \frac{2}{3^4}\right) + \left(\frac{1}{3^5} + \frac{2}{3^6}\right) + \ldots \\ &= \frac{5}{3^2} + \frac{5}{3^4} + \frac{5}{3^6} + \ldots \\ &= \sum_{n=1}^{\infty} \frac{5}{9^n}. \end{aligned}$$

This last series is a convergent geometric series with $a = \frac{5}{9}$ and $r = \frac{1}{9}$. Therefore

$$x = \frac{\frac{5}{9}}{1 - \frac{1}{9}} = \frac{5}{8}.$$

Exercise 10.24. Describe the rational number with base 4 representation $0.\overline{301}_4$.

We will now go the other direction; beginning with a rational number, we will obtain a base b representation for it. Our method will be altogether analogous to the method used to obtain the base 10 representation of a real number. We let $y_0 = 0$, and then for each $n \geq 1$, we let y_n be the largest integer with $y_n \leq b^n x$. The n-th digit in the base b representation of x is then $a_n = y_n - by_{n-1}$. The proof that, for a_n defined in this way, $x = \sum_{n=1}^{\infty} a_n b^{-n}$ is virtually identical to our proof for base 10 representations and we omit it. We instead illustrate the use of the algorithm in a concrete example. In the programming project at the end of the chapter, you will have the opportunity to write code to execute this algorithm.

Example 10.11. We obtain the base 3 representation of $x = \frac{1}{4}$.

For $n = 1$, we note that $0 \leq 3 \cdot \frac{1}{4}$ but $1 > 3 \cdot \frac{1}{4}$. Thus $y_1 = 0$ and $a_1 = 0$. For $n = 2$, note that $2 \leq 3^2 \cdot \frac{1}{4}$ but $3 > 3^2 \cdot \frac{1}{4}$. Thus $y_2 = 2$ and $a_2 = y_2 - 3y_1 = 2$. For $n = 3$, note that $6 \leq 3^3 \cdot \frac{1}{4}$ but $7 > 3^3 \cdot \frac{1}{4}$. Thus $y_3 = 6$ and $a_3 = y_3 - 3y_2 = 6 - 3 \cdot 2 = 0$. For $n = 4$, note that $20 \leq 3^4 \cdot \frac{1}{4}$ but $21 > 3^4 \cdot \frac{1}{4}$. Thus $y_4 = 20$ and $a_4 = y_4 - 3y_3 = 20 - 3 \cdot 6 = 2$. So far we have obtained the approximation 0.0202_3 for x. One can easily show that in fact $x = 0.\overline{02}_3$.

Exercise 10.25. Find the base 2 representation of $\frac{2}{3}$.

10.5 Problems

1. Determine whether each of the following series is convergent or divergent. You need not find the value of a convergent series.

   (a) $\displaystyle\sum_{n=0}^{\infty} \frac{2^n}{n!}$.

   (b) $\displaystyle\sum_{n=1}^{\infty} \frac{n}{(n+1)^2}$.

   (c) $\displaystyle\sum_{n=1}^{\infty} \frac{n!}{n^n}$.

   (d) $\displaystyle\sum_{n=1}^{\infty} \frac{n^n}{n!}$.

   (e) $\displaystyle\sum_{n=0}^{\infty} \frac{\sqrt{n}}{n^2+4}$.

2. For which real numbers x does $\displaystyle\sum_{n=0}^{\infty} \frac{(x-2)^n}{2^n n^3}$ converge?

3. Let $\{a_n\}$ and $\{b_n\}$ be two sequences of real numbers.

   (a) Suppose $\sum(a_n + b_n)$ converges. Must $\sum a_n$ and $\sum b_n$ converge?

(b) Suppose $\sum a_n b_n$ converges. Must $\sum a_n$ and $\sum b_n$ converge?

4. The *Cantor middle-thirds set* is constructed as follows: Let $C_0 = [0,1]$. Then, for $n > 0$, obtain the set C_n from C_{n-1} by deleting from each closed interval making up C_{n-1} an open interval that is its "middle third." Thus to create C_1, we remove $(\frac{1}{3}, \frac{2}{3})$ from $[0,1]$ and so $C_1 = [0, \frac{1}{3}] \cup [\frac{2}{3}, 1]$. The Cantor set is then $\bigcap_{n=0}^{\infty} C_n$.

   (a) Find C_2, C_3, C_4.

   (b) Let L_n be the sum of the lengths of the intervals removed to obtain C_n from C_{n-1}. (Thus L_0 is not defined, $L_1 = \frac{1}{3}$, $L_2 = \frac{2}{9}$, etc.) Find a simple formula for L_n.

   (c) Find $\sum_{n=1}^{\infty} L_n$. This number represents the sum of the lengths of all intervals removed to form the Cantor set.

5. Prove the Limit Comparison Test: Let $\{a_n\}$ and $\{b_n\}$ be sequences of positive real numbers. Suppose $\lim_{n\to\infty} a_n/b_n = L$, where L is a non-zero real number. Then $\sum a_n$ converges if and only if $\sum b_n$ converges.

6. Let p and q be polynomials and suppose that neither is the zero polynomial. Consider $\sum_{n=n_0}^{\infty} \frac{p(n)}{q(n)}$ where n_0 is large enough that $q(n) \neq 0$ for all $n \geq n_0$. Such an n_0 exists because q has only finitely many zeros and we can take n_0 to be a natural number greater than the maximum of the zeros of q. State and prove a theorem that relates the convergence or divergence of the series $\sum_{n=n_0}^{\infty} \frac{p(n)}{q(n)}$ to the degrees of p and q. (Remark: Although many approaches are possible, you may find the limit comparison test from the previous exercise to be particularly helpful.)

7. You may have been taught in elementary or middle school to use the following procedure to convert a repeating decimal to a rational number. Suppose $x = 0.\overline{a_1 \ldots a_N}$. Then $10^N x = a_1 \ldots a_N.\overline{a_1 \ldots a_N}$. Subtract the former from the latter, and solve for x.

   (a) Use this procedure to write $0.\overline{4}$ and $0.\overline{123}$ as ratios of integers.

   (b) We claim the above procedure relies on a number of propositions about infinite series. Write a more detailed description of the procedure that justifies each step by stating explicitly what results about series are used.

8. Let $b \geq 2$ be a natural number. Each of the following gives the base b expansion of a rational number. Identify that rational number by writing it as a ratio of integers.

   (a) $0.\overline{1}_b$.

   (b) $0.\overline{(b-1)}_b$.

9. Describe as precisely as you can all real numbers x with $0 \leq x \leq 1$ whose base 2 representation is not unique.

10. In this problem we consider series $\sum a_n$ of *complex* numbers.

    (a) Does Proposition 10.4 apply to series of complex numbers? Prove or give a counterexample.

    (b) Determine whether $\sum_{n=1}^{\infty} \frac{(i)^n}{n^2}$ converges.

Programming Project. Write a program that takes as input a base b and the numerator and denominator of a rational number x between 0 and 1 and returns the first 8 digits in the base b representation of x. We suggest that you use the algorithm described in the text. If you do so, you will have to do comparisons to find the numbers y_n. Try to keep the number of comparisons necessary as small as you can by using the fact that $y_n \geq b y_{n-1}$.

Chapter 11

The Structure of the Real Line

We now know rather a lot about the real number system; we know that $\mathbb{R}$ is a complete ordered field. This knowledge has allowed us to develop a satisfying theory of limits of sequences because we have a nice necessary and sufficient condition for convergence, namely the Cauchy criterion. This theory has, in turn, given us the tools we need to study infinite series of real numbers and representations of real numbers.

In this chapter, we explore additional properties of $\mathbb{R}$ and its subsets that will be important as we develop a theory of functions from $\mathbb{R}$ to $\mathbb{R}$. In particular, in the next chapter, we will talk about limits of functions and properties of continuous functions. For many of our definitions and theorems, the exact nature of the set on which the function is defined will be important. To illustrate this point and to motivate the definitions in this chapter, let's consider a familiar theorem from calculus. This theorem plays an important role in the procedure we use in optimization problems.

Theorem 11.1. *Let $f\colon [a,b] \to \mathbb{R}$ be a continuous function. Then f attains a maximum and a minimum value, i.e., there exists $c \in [a,b]$ such that $f(x) \leq f(c)$ for all $x \in [a,b]$ and there exists $d \in [a,b]$ such that $f(d) \leq f(x)$ for all $x \in [a,b]$.*

In Theorem 11.1, is the form of the set on which f is continuous important? Would the same theorem hold if, for example, the hypothesis were that f is continuous on the open interval (a,b)? The answer is No. Consider, for example, the function $f(x) = \frac{1}{x}$. We will show in the next chapter that f is continuous on (for example) $(0,1)$. This function does not attain a maximum value because it is unbounded. It also does not attain a minimum value; although it is bounded below, the infimum of its image set is 1, but there is no $d \in (0,1)$ for which $f(d) = 1$.

What property or properties does the set $[a,b]$ have that the set $(0,1)$ does not have? In this chapter we explore this question and other related questions about the structure of subsets of the real line.

11.1 Basic Notions from Topology

11.1.1 Open and closed sets

In earlier courses, you have met open and closed *intervals*. Recall, if $a < b$, $(a, b) = \{x \in \mathbb{R} : a < x < b\}$ and $[a, b] = \{x \in \mathbb{R} : a \leq x \leq b\}$. The first set, the open interval, has the property that every point in the interval is completely surrounded by points of the interval. The second set does not have this property; there are two points (a and b) for which every little interval around them contains both points in the interval and points not in the interval. The next definition generalizes the notion of an open interval and also makes precise what it means for every point in a set to be completely surrounded by other points of the set.

Definition 11.1. Let $U \subseteq \mathbb{R}$. U is **open** if, for every $x \in U$, there exists $\varepsilon > 0$ such that $(x - \varepsilon, x + \varepsilon) \subseteq U$.

It will not surprise you to hear that open intervals are examples of open sets.

Proposition 11.1. *Suppose $a < b$. Then (a, b) is open.*

Proof. Let $x \in (a, b)$, so that $a < x < b$. Let $\varepsilon = \min\{b - x, x - a\}$. We claim that $(x - \varepsilon, x + \varepsilon) \subseteq (a, b)$. Indeed, let $y \in (x - \varepsilon, x + \varepsilon)$. We must show that $y \in (a, b)$. Observe,

$$\begin{aligned} y &> x - \varepsilon \\ &= x - \min\{b - x, x - a\} \\ &\geq x - (x - a) = a. \end{aligned}$$

Similarly,

$$\begin{aligned} y &< x + \varepsilon \\ &= x + \min\{b - x, x - a\} \\ &\leq x + (b - x) = b. \end{aligned}$$

These inequalities show that $y \in (a, b)$. Because y was arbitrary, $(x - \varepsilon, x + \varepsilon) \subseteq (a, b)$. □

As simple as this proposition is, it illustrates the approach we use to prove that a set is open. As always, our definition tells us exactly what to do. We take an arbitrary element x out of the set and show that we can find a positive ε so that the entire interval $(x - \varepsilon, x + \varepsilon)$ remains in the set. In order to prove that a set U is *not* open, of course we start with the negation of the statement defining openness. Thus U is not open if there exists $x \in U$ such that, for every $\varepsilon > 0$, $(x - \varepsilon, x + \varepsilon)$ contains a point that is not in U.

Example 11.1. We show that $[0, 1) = \{x \in \mathbb{R} : 0 \leq x < 1\}$ is not open. Consider the element 0 of the set. Let $\varepsilon > 0$ be given. Then $(0 - \varepsilon, 0 + \varepsilon) = (-\varepsilon, \varepsilon)$ contains $y = -\frac{\varepsilon}{2}$. Because $y < 0$, $y \notin [0, 1)$. Thus $(-\varepsilon, \varepsilon)$ is not contained in $[0, 1)$ for any ε. We conclude that $[0, 1)$ is not open.

Exercise 11.1. Show that both $\emptyset$ and $\mathbb{R}$ are open subsets of $\mathbb{R}$.

Exercise 11.2. Show that both $(-\infty, a) = \{x \in \mathbb{R} : x < a\}$ and $(a, \infty) = \{x \in \mathbb{R} : x > a\}$ are open sets.

Exercise 11.3. Let $x \in \mathbb{R}$. Show that $\{x\}$ is not open.

It is important to know what set operations preserve the class of open sets.

Proposition 11.2. *(a) Let $\{U_\alpha : \alpha \in \mathcal{A}\}$ be an arbitrary collection of open sets. Then $\bigcup_{\alpha \in \mathcal{A}} U_\alpha$ is open.*

(b) Let $\{U_j : 1 \leq j \leq n\}$ be a finite collection of open sets. Then $\bigcap_{j=1}^{n} U_j$ is open.

Proof. For part (a), let $x \in \bigcup_{\alpha \in \mathcal{A}} U_\alpha$. Then there exists $\alpha_0 \in \mathcal{A}$ such that $x \in U_{\alpha_0}$. Because U_{α_0} is open, there exists $\varepsilon > 0$ such that $(x - \varepsilon, x + \varepsilon) \subseteq U_{\alpha_0}$. Because $U_{\alpha_0} \subseteq \bigcup_{\alpha \in \mathcal{A}} U_\alpha$, $(x - \varepsilon, x + \varepsilon) \subseteq \bigcup_{\alpha \in \mathcal{A}} U_\alpha$. Thus $\bigcup_{\alpha \in \mathcal{A}} U_\alpha$ is open.

For part (b), let $x \in \bigcap_{j=1}^{n} U_j$, so that $x \in U_j$ for all $1 \leq j \leq n$. For each j, because U_j is open, there exists $\varepsilon_j > 0$ such that $(x - \varepsilon_j, x + \varepsilon_j) \subseteq U_j$. Set $\varepsilon = \min\{\varepsilon_j : 1 \leq j \leq n\}$ and note that ε is positive. Because $\varepsilon \leq \varepsilon_j$, $(x - \varepsilon, x + \varepsilon) \subseteq U_j$ for all j and so $(x - \varepsilon, x + \varepsilon) \subseteq \bigcap_{j=1}^{n} U_j$. Thus $\bigcap_{j=1}^{n} U_j$ is open. $\square$

Suppose we try to extend our proof of (b) to show that an intersection of countably many open sets is open. Thus suppose $x \in \bigcap_{j=1}^{\infty} U_j$ and each U_j is open. It is still true that for each j, there exists $\varepsilon_j > 0$ such that $(x - \varepsilon_j, x + \varepsilon_j) \subseteq U_j$. However, if we continue to the next step of the proof, we encounter a problem. The set $\{\varepsilon_j : j \in \mathbb{N}\}$ may not have a minimum element, and its infimum may be zero. If it is, we will not have obtained a positive ε for which $(x - \varepsilon, x + \varepsilon)$ is contained in every U_j. Our inability to make *this* proof method work for an infinite collection of sets does not allow us to conclude that the result is false, but it is evidence pointing in that direction that suggests where we might look for a counterexample. In fact, an easy counterexample exists.

Example 11.2. Let $U_j = \left(-\frac{1}{j}, \frac{1}{j}\right)$. Each U_j is open because it is an open interval, but $\bigcap_{j=1}^{\infty} U_j = \{0\}$, which is not open by Exercise 11.3.

With this understanding of open sets, we may now turn our attention to closed sets.

Definition 11.2. Let $E \subseteq \mathbb{R}$. E is **closed** if E^c is open.

Just as our prototypical example of an open set is an open interval, our prototypical example of a closed set is a closed interval.

Proposition 11.3. *Let $a, b \in \mathbb{R}$ with $a < b$. Then $[a, b]$ is closed.*

Proof. Our definition requires us to show that $[a, b]^c$ is open. Observe,

$$[a, b]^c = (-\infty, a) \cup (b, \infty).$$

By Exercise 11.2, each set on the right is open, and by Proposition 11.2, their union is open. Because $[a, b]^c$ is open, by definition, $[a, b]$ is closed. $\square$

We should also give an example to illustrate how we show that a set is not closed. By definition, in order to show that a set E is not closed, we must show that its complement E^c is not open. Such arguments are now familiar.

Example 11.3. We show that $[0, 1)$ is not closed. We must consider its complement

$$[0, 1)^c = (-\infty, 0) \cup [1, \infty).$$

For convenience we name this set S. Observe that 1 is an element of S. Let $\varepsilon > 0$ be given. Let $y = \max\{1 - \frac{\varepsilon}{2}, 0\}$. Then $y \in (1 - \varepsilon, 1 + \varepsilon)$ because

$$\begin{aligned} 1 &> y \\ &\geq 1 - \frac{\varepsilon}{2} \\ &> 1 - \varepsilon. \end{aligned}$$

Also, $y \notin S$ because $0 \leq y < 1$. Thus $(1 - \varepsilon, 1 + \varepsilon)$ contains a point that is not in S. Because ε was arbitrary, S is not open. Therefore $[0, 1)$ is not closed.

Exercise 11.4. Show that both $\emptyset$ and $\mathbb{R}$ are closed.

Exercise 11.5. Show that any finite set $\{x_1, \ldots, x_n\}$ is closed.

In English, *open* and *closed* are often considered to be opposites. In mathematics, however, they are not. We have seen that there are sets like $[0, 1)$ that are neither open nor closed, and there are sets like $\emptyset$ and $\mathbb{R}$ that are both open and closed. Thus we see again the importance of heeding our definitions and putting aside common usage.

Our last proposition in this section is the analogue of Proposition 11.2 for closed sets. We leave the proof to the reader.

Proposition 11.4. *(a) Let $\{E_\alpha : \alpha \in \mathcal{A}\}$ be an arbitrary collection of closed sets. Then $\bigcap_{\alpha \in \mathcal{A}} E_\alpha$ is closed.*

(b) Let $\{E_j : 1 \leq j \leq n\}$ be a finite collection of closed sets. Then $\bigcup_{j=1}^n E_j$ is closed.

Exercise 11.6. Prove Proposition 11.4. You will want to recall De Morgan's laws.

11.1.2 Accumulation points of sets

In the previous subsection, we defined a closed set to be a set whose complement is open. We give here a second characterization involving the notion of an accumulation point of a set.

Definition 11.3. Let $A \subseteq \mathbb{R}$ and let x be an element of $\mathbb{R}$ not necessarily in A. Then x is an **accumulation point** of A if, for every $\varepsilon > 0$, $(x - \varepsilon, x + \varepsilon)$ contains an element of A different from x.

Example 11.4. To illustrate the definition, let's find all accumulation points of the set $A = (0, 1)$. We observe first that, in this case, any element of A is an accumulation point of A. Indeed, let $x \in A$ and let $\varepsilon > 0$ be given. If $x - \varepsilon < 0$, then $y = \frac{x}{2}$ satisfies

$$x - \varepsilon < 0 < y < x$$

and thus y is an element of A different from x in $(x - \varepsilon, x + \varepsilon)$. If $x - \varepsilon \geq 0$, then $z = x - \frac{\varepsilon}{2}$ satisfies

$$0 \leq x - \varepsilon < z < x$$

and thus z is an element of A different from x in $(x - \varepsilon, x + \varepsilon)$.

Next we claim that 0 and 1 are accumulation points of A. Let $\varepsilon > 0$ be given. Then $y = \min\{\frac{1}{2}, \frac{\varepsilon}{2}\}$ is an element of A in $(-\varepsilon, \varepsilon)$ and $z = \max\{\frac{1}{2}, 1 - \frac{\varepsilon}{2}\}$ is an element of A in $(1 - \varepsilon, 1 + \varepsilon)$. These claims are established.

Finally, we claim that any point in $U = (-\infty, 0) \cup (1, \infty)$ is not an accumulation point of A. Note that U is open and disjoint from A. Thus for any $x \in U$, there exists $\varepsilon > 0$ such that $(x - \varepsilon, x + \varepsilon) \subseteq U$ and hence contains no points of A. Thus x is not an accumulation point.

The intuition is supposed to be that an accumulation point of a set may or may not be in the set, but it has lots of elements of the set close to it. This intuition is strengthened by the following proposition characterizing accumulation points.

Proposition 11.5. *Let $A \subseteq \mathbb{R}$ and let x be an element of $\mathbb{R}$ not necessarily in A. Then x is an accumulation point of A if and only if, for every $\varepsilon > 0$, $(x - \varepsilon, x + \varepsilon)$ contains infinitely many elements of A.*

Proof. Because we wish to prove a biconditional statement, we must prove two implications.

Suppose first that x has the property that, for every $\varepsilon > 0$, $(x - \varepsilon, x + \varepsilon)$ contains infinitely many elements of A. It then clearly contains an element of A different from x, and so x is indeed an accumulation point.

Conversely, suppose x is an accumulation point of A. Let $\varepsilon > 0$ be given. Then $(x - \varepsilon, x + \varepsilon)$ contains an element a_1 of A such that $a_1 \neq x$. Now set $\varepsilon_1 = |x - a_1|$ and note that $0 < \varepsilon_1 < \varepsilon$. Consider $(x - \varepsilon_1, x + \varepsilon_1)$. This interval contains an element a_2 of A different from x. Furthermore, $a_2 \neq a_1$ because a_1 is not an element of $(x - \varepsilon_1, x + \varepsilon_1)$. Proceed inductively; having selected distinct elements $a_1, \ldots, a_n$ of A different from x, set $\varepsilon_n = |x - a_n|$.

Note that $0 < \varepsilon_n < \varepsilon_{n-1} < \ldots < \varepsilon_1 < \varepsilon$. Then take a_{n+1} to be any element of A in $(x - \varepsilon_n, x + \varepsilon_n)$ different from x. Because no a_j for $1 \le j \le n$ is in $(x - \varepsilon_n, x + \varepsilon_n)$, a_{n+1} is also unequal to a_j for all $1 \le j \le n$. In this manner we obtain an infinite number of elements of A in the interval $(x - \varepsilon, x + \varepsilon)$, as desired. □

Exercise 11.7. Find all accumulation points of the sets $A = \{\frac{1}{n} : n \in \mathbb{N}\}$ and $B = [-1, 1]$.

The idea from the second half of the proof of Proposition 11.5 can be modified to establish a connection between accumulation points of sets and limits of sequences in the set.

Proposition 11.6. *If x is an accumulation point of A, then there exists a sequence $\{a_n\}$ consisting of elements of A such that $\lim_{n\to\infty} a_n = x$.*

Proof. For each n, consider $(x - 1/n, x + 1/n)$. Because x is an accumulation point of A, there exists $a_n \in (x - 1/n, x + 1/n) \cap A$. We claim $\{a_n\}$ has limit x. Indeed, let $\varepsilon > 0$ be given. Then there exists N such that $\frac{1}{N} < \varepsilon$. For all $n \ge N$,

$$x - \varepsilon < x - \frac{1}{N} < x - \frac{1}{n} < a_n < x + \frac{1}{n} < x + \frac{1}{N} < x + \varepsilon,$$

i.e., for all $n \ge N$, $|a_n - x| < \varepsilon$. □

Exercise 11.8. Modify the proof of Proposition 11.6 to show that, if x is an accumulation point of A, there exists a sequence $\{a_n\}$ *of distinct elements of A* such that $\lim_{n\to\infty} a_n = x$.

The next theorem is the main goal of this subsection. It makes the connection between accumulation points and closed sets.

Theorem 11.2. *$E \subseteq \mathbb{R}$ is closed if and only if E contains all its accumulation points.*

Proof. Suppose first that E is closed, so that E^c is open. We prove that if x is an accumulation point of E, then $x \in E$ by proving the contrapositive, i.e., if x is not in E, then x is not an accumulation point.

If $x \notin E$, then $x \in E^c$. Because E^c is open, there exists $\varepsilon > 0$ such that $(x - \varepsilon, x + \varepsilon) \subseteq E^c$. This set thus contains no points of E and hence x is not an accumulation point.

Conversely, suppose E contains all its accumulation points. We must show E is closed, i.e., that E^c is open. Take $x \in E^c$. Because E contains all its accumulation points, x is not an accumulation point of E. Thus there exists $\varepsilon > 0$ such that $(x - \varepsilon, x + \varepsilon)$ contains no element of E different from x. Because x itself is not in E, we have $(x - \varepsilon, x + \varepsilon) \subseteq E^c$. Thus E^c is indeed open. □

Exercise 11.9. Prove that $\{\frac{1}{n} : n \in \mathbb{N}\} \cup \{0\}$ is closed.

Exercise 11.10. Consider the finite set $A = \{a_1, \ldots, a_n\}$. What are the accumulation points of A? Prove that A is closed by arguing that it contains all its accumulation points.

11.2 Compact sets

Consider again Theorem 11.1. In that theorem, the hypothesis on f is that it is continuous on $[a, b]$. We now know that $[a, b]$ is an example of a closed set. Is the theorem still true if we replace $[a, b]$ by an arbitrary closed set? No. We know, for example, that $\mathbb{R}$ is closed. The function $f\colon \mathbb{R} \to \mathbb{R}$ defined by $f(x) = x^2$ is an example of a continuous function that is unbounded on $\mathbb{R}$, hence does not attain a maximum value. Thus requiring the domain to be closed is not enough; we have not yet found the right topological property. It turns out that we should require the domain to be both closed *and bounded*. The sets with both these properties are the *compact sets*, though it will not be immediately clear from our definition of compact sets that they have these properties.

11.2.1 Subsequences and limit points

Before we give the definition of a compact set, we must talk briefly about subsequences of sequences of real numbers.

Definition 11.4. Let $a = \{a_n\}$ be a sequence of real numbers. Sequence $b = \{b_k\}$ is a **subsequence** of a if there exists a strictly increasing function $n\colon \mathbb{N}_0 \to \mathbb{N}_0$ such that $b_k = a_{n(k)}$ for all k. We call $k \mapsto n(k)$ the **subsequence selection function**. We often write $\{a_{n_k}\}$ for a subsequence instead of $\{a_{n(k)}\}$.

As always, we look at an example to clarify the definition.

Example 11.5. Consider the sequence a with n-th term $a_n = \frac{1}{n+1}$. If we define $n\colon \mathbb{N}_0 \to \mathbb{N}_0$ by $n(k) = 2k$, we obtain the subsequence

$$\begin{aligned}\{a_{n(0)}, a_{n(1)}, a_{n(2)}, \ldots, a_{n(k)}, \ldots\} &= \{a_0, a_2, a_4, \ldots, a_{2k}, \ldots\} \\ &= \left\{1, \frac{1}{3}, \frac{1}{5}, \ldots, \frac{1}{2k+1}, \ldots\right\}.\end{aligned}$$

If instead we take $n(k) = k^2$, we obtain the subsequence with $b_k = \frac{1}{k^2+1}$. Recalling that sequences are themselves functions on $\mathbb{N}_0$, we observe that all we are really doing to get sequence b is taking a and composing it with a strictly increasing function n on $\mathbb{N}_0$ to obtain $a \circ n\colon \mathbb{N}_0 \to \mathbb{R}$.

Example 11.6. For a more interesting example, we consider sequence a defined by $a_n = (-1)^n \frac{n}{n+1}$. This sequence is divergent, for the terms with even index are approaching 1 and the terms with odd index are approaching -1. We can express this idea more precisely by considering two subsequences. For the first, let $n(k) = 2k$. We obtain the subsequence b with k-th term

$$b_k = a_{n(k)} = (-1)^{2k} \frac{2k}{2k+1} = \frac{2k}{2k+1}.$$

It is easy to see that b is convergent, with limit 1. For the second subsequence, let $n(k) = 2k + 1$. We obtain the subsequence c with k-th term

$$c_k = a_{n(k)} = (-1)^{2k+1} \frac{2k+1}{2k+2} = -\frac{2k+1}{2k+2}.$$

It is easy to see that c is convergent with limit -1.

Example 11.6 shows that a divergent sequence may have convergent subsequences. We are thus motivated to make a definition.

Definition 11.5. Let $a = \{a_n\}$ be a sequence of real numbers. A real number L is a **limit point** of a if there exists a subsequence of a with limit L.

The sequence in Example 11.6 has limit points 1 and -1. How many limit points can a sequence have? To explore this question, we consider two more examples. The first is elementary and so we leave it as an exercise. The second is a little more interesting.

Exercise 11.11. Let a be the sequence with n-th term $a_n = n \bmod 3$. In other words,

$$a = \{0, 1, 2, 0, 1, 2, 0, 1, 2, \ldots\}.$$

What are the limit points of a? For each limit point, describe a subsequence $\{a_{n(k)}\}$ with that limit by giving an explicit subsequence selection function $k \mapsto n(k)$.

Example 11.7. It is possible for a sequence to have infinitely many limit points. For example,

$$a = \{1, 1, 2, 1, 2, 3, 1, 2, 3, 4, 1, 2, 3, 4, 5, \ldots\}$$

has every natural number as a limit point.

We prove a simple but important result about limit points of convergent sequences.

Proposition 11.7. *Let $a = \{a_n\}$ be a convergent sequence with limit L. Then every subsequence has limit L. Consequently, a convergent sequence has a single limit point.*

Proof. Let $\{b_k\} = \{a_{n_k}\}$ be any subsequence of a and let $\varepsilon > 0$ be given. Because the full sequence a converges to L, for this $\varepsilon > 0$, there exists N such that, for all $n \geq N$, $|a_n - L| < \varepsilon$. Because $k \mapsto n_k$ is strictly increasing, if $k \geq N$, $n_k \geq N$ and so

$$|b_k - L| = |a_{n_k} - L| < \varepsilon.$$

Thus $\{b_k\}$ has limit L as well. □

11.2.2 First definition of compactness

We are now ready to give the definition of a compact set.

Definition 11.6. A set $K \subseteq \mathbb{R}$ is **compact** if every sequence of elements of K has a limit point in K.

What kinds of sets are compact? To begin with, sets with only finitely many elements are compact.

Proposition 11.8. *Let $K = \{x_1, \ldots, x_M\}$ be a finite subset of $\mathbb{R}$. Then K is compact.*

Proof. Let a be a sequence indexed by $\mathbb{N}$ such that $a_n \in K$ for all n. We must show that some subsequence of a converges to an element of K. Because K is finite, some element of K must be equal to a_n for infinitely many n. Let x_j be such an element of K. We will obtain a subsequence of a with limit x_j. Let $A = \{n \in \mathbb{N} : a_n = x_j\}$. The set A is an infinite (hence non-empty) subset of $\mathbb{N}$. By the well-ordering principle for $\mathbb{N}$, A has a smallest element n_1. Proceed inductively; having obtained $n_1 < n_2 < \ldots < n_k$ for $k \geq 1$, we obtain n_{k+1} as follows: The set $A \setminus \{n_1, \ldots, n_k\}$ is still an infinite (hence non-empty) subset of $\mathbb{N}$ and so it has a smallest element n_{k+1}. Clearly $n_{k+1} > n_k$, for otherwise, at step k, n_k would not have been the smallest element of $A \setminus \{n_1, \ldots, n_{k-1}\}$. Define a sequence b by setting $b_k = a_{n_k}$. Because $k \mapsto n_k$ is strictly increasing, b is a subsequence of a. Because b is a constant sequence with every term equal to x_j, it is clearly convergent to the element x_j of K. Thus K is compact. $\square$

Exercise 11.12. Negate the definition of compactness and use the negation to show that $\mathbb{Z}$ is not compact.

Exercise 11.13. Show that $(0, 1)$ is not compact.

Another canonical example of a compact set is a closed and bounded interval.

Proposition 11.9. *Let $a, b \in \mathbb{R}$ with $a < b$. Then $K = [a, b]$ is compact.*

Proof. Let $\{x_n\}$ be a sequence in $[a, b]$ indexed by $\mathbb{N}$. We will find a convergent subsequence of $\{x_n\}$ using a divide-and-conquer approach. Set $c_1 = \frac{a+b}{2}$. One of the sets $[a, c_1]$ or $[c_1, b]$ contains infinitely many terms of the sequence $\{x_n\}$. Let $I_1 = [a_1, b_1]$ be one of the sets with this property and let x_{n_1} be a term in the sequence $\{x_n\}$ in I_1. We proceed inductively to define the sets I_k and the terms x_{n_k}; having obtained closed intervals $I_{k-1} \subseteq \ldots \subseteq I_1$ and terms $x_{n_1}, \ldots, x_{n_{k-1}}$ of $\{x_n\}$ with $n_1 < \ldots < n_{k-1}$, consider the midpoint $c_k = \frac{a_{k-1}+b_{k-1}}{2}$ of $I_{k-1} = [a_{k-1}, b_{k-1}]$. Because I_{k-1} contains infinitely many terms of the sequence $\{x_n\}$, one of the subintervals $[a_{k-1}, c_k]$ or $[c_k, b_{k-1}]$ does as well. Let $I_k = [a_k, b_k]$ be one of these intervals having this property and let x_{n_k} be a term in the sequence $\{x_n\}$ that is in I_k and for which $n_k > n_{k-1}$.

In this manner we have constructed a subsequence $\{x_{n_k}\}$ of $\{x_n\}$. We claim it is convergent to an element of $[a, b]$. We prove this claim by proving that the sequence $\{x_{n_k}\}$ is Cauchy. Let $\varepsilon > 0$ be given. We must show that there exists J such that for all $j, k \geq J$, $|x_{n_j} - x_{n_k}| < \varepsilon$. Assume without loss of generality that $k > j$ and observe that $I_k \subseteq I_j$ for all k. Observe also that because $b_j - a_j = \frac{1}{2^j}(b - a)$, if $x_{n_k}, x_{n_j} \in I_j$,

$$|x_{n_k} - x_{n_j}| \leq \frac{1}{2^j}(b - a).$$

Now, $\frac{1}{2^j}(b - a) < \varepsilon$ if $j > \log_2\left(\frac{b-a}{\varepsilon}\right)$. Thus if we take J to be any natural number larger than $\log_2\left(\frac{b-a}{\varepsilon}\right)$, then for all $k > j \geq J$, $|x_{n_k} - x_{n_j}| < \varepsilon$. We conclude that $\{x_{n_k}\}$ is a subsequence of $\{x_n\}$ that converges to some y. Because for all $k \geq J$, $x_{n_k} \in I_J$, for all such k,

$$a_J \leq x_{n_k} \leq b_J.$$

Because non-strict inequalities are preserved when taking limits, $a_J \leq y \leq b_J$. In particular, $y \in [a, b]$. Because $\{x_n\}$ was an arbitrary sequence in $[a, b]$, we have shown that every sequence in $[a, b]$ has a limit point in $[a, b]$. Thus, by definition, $[a, b]$ is compact. □

At the beginning of the section, we indicated that in Theorem 11.1, the special property of the domain of f guaranteeing that a function attains its maximum and minimum is that it is closed and bounded, but then we defined compactness in terms of sequences and subsequences. We bring these two notions together in a general theorem characterizing compact sets.

Theorem 11.3. *$K \subseteq \mathbb{R}$ is compact if and only if it is closed and bounded.*

Proof. We begin by proving that if K is compact, then it is closed and bounded. We accomplish this goal by proving the contrapositive. Thus we first suppose K is not bounded and show that K is not compact. Indeed, if K is not bounded, then given $n \in \mathbb{N}$ there exists $a_n \in K$ such that $|a_n| > n$. Form the sequence $\{a_n\}$. Because every subsequence of $\{a_n\}$ is also clearly unbounded, $\{a_n\}$ has no convergent subsequences, hence no limit points. Thus K is not compact.

Suppose next that K is not closed. Then by Theorem 11.2, there exists an accumulation point x of K such that $x \notin K$. By Proposition 11.6, there exists a sequence $\{a_n\}$ such that $a_n \in K$ and $\lim_{n\to\infty} a_n = x$. Because $\{a_n\}$ is convergent, x is its only limit point. Thus this sequence is an example of a sequence in K with no limit point in K. Thus K is not compact.

Next we must show that, if K is closed and bounded, then K is compact. The structure of this proof is virtually identical to that of our proof of Proposition 11.9. Because K is bounded, there exists $M > 0$ such that $K \subseteq [-M, M]$. Let $\{x_n\}$ be a sequence in K indexed by $\mathbb{N}$. Set $c_1 = \frac{-M+M}{2} = 0$. Because K is contained in their union, one of the intervals $[-M, c_1]$ or $[c_1, M]$ contains infinitely many terms of the sequence $\{x_n\}$. Let $I_1 = [a_1, b_1]$ be one such interval with this property and let x_{n_1} be a term in the sequence $\{x_n\}$ in I_1. We proceed inductively; having defined closed intervals $I_{k-1} \subseteq \ldots \subseteq I_1$ and terms $x_{n_1}, \ldots, x_{n_{k-1}}$ of $\{x_n\}$ with $n_1 < \ldots < n_{k-1}$, consider the midpoint $c_k = \frac{a_{k-1}+b_{k-1}}{2}$ of $I_{k-1} = [a_{k-1}, b_{k-1}]$. Because I_{k-1} contains infinitely many terms of the sequence $\{x_n\}$, one of the subintervals $[a_{k-1}, c_k]$ or $[c_k, b_{k-1}]$ does as well. Let $I_k = [a_k, b_k]$ be one such interval and let x_{n_k} be a term in the sequence $\{x_n\}$ that is in I_k and for which $n_k > n_{k-1}$. We have constructed a subsequence $\{x_{n_k}\}$ of $\{x_n\}$. We claim it converges to an element of K. We prove this claim by first proving that $\{x_{n_k}\}$ is Cauchy. Let $\varepsilon > 0$ be given. We must show that there exists J such that for all $j, k \geq J$, $|x_{n_j} - x_{n_k}| < \varepsilon$. Assume without loss of generality that $k > j$ and observe that $I_k \subseteq I_j$ for all k.

Observe also that because $b_j - a_j = \frac{2M}{2^j}$, if $x_{n_k}, x_{n_j} \in I_j$,

$$|x_{n_k} - x_{n_j}| \le \frac{M}{2^{j-1}}.$$

Now, $\frac{M}{2^{j-1}} < \varepsilon$ if $j > \log_2\left(\frac{M}{\varepsilon}\right) + 1$. Thus if we take J to be any natural number larger than $\log_2\left(\frac{M}{\varepsilon}\right) + 1$, then for all $k > j \ge J$, $|x_{n_k} - x_{n_j}| < \varepsilon$. We conclude that $\{x_{n_k}\}$ is a subsequence of $\{x_n\}$ that is Cauchy, hence converges to some y. We claim that $y \in K$. If not, $y \in K^c$. Because K is closed, K^c is open, and so there exists $\varepsilon > 0$ such that $(y - \varepsilon, y + \varepsilon)$ is contained in K^c. This contradicts the fact that the x_n are in K and that $\lim_{k\to\infty} x_{n_k} = y$. Thus $y \in K$ and K is compact. □

11.2.3 The Heine–Borel property

The definition of compactness given in the last subsection is the right one for a first course in analysis, and it is the right one for this book because of the weight we give to infinite sequences. There is, however, another way to think about compactness that is entirely topological; it makes no reference to sequences but instead refers only to collections of open sets. We include it here for completeness. We use it in the second half of this chapter but otherwise avoid it.

Definition 11.7. Let E be a subset of $\mathbb{R}$. A collection $\mathcal{C} = \{U_\alpha : \alpha \in \mathcal{A}\}$ is an **open cover** for E if each U_α is an open subset of $\mathbb{R}$ and $E \subseteq \bigcup_{\alpha\in\mathcal{A}} U_\alpha$. A collection $\{U_j : 1 \le j \le n\}$ is a **finite subcover** of E if each $U_j \in \mathcal{C}$ and $E \subseteq \bigcup_{j=1}^{n} U_j$.

Definition 11.8. Let E be a subset of $\mathbb{R}$. Then E has the **Heine–Borel property** if every open cover of E has a finite subcover.

These definitions take some getting used to and so we consider some examples.

Example 11.8. Let $E = \{\frac{1}{n} : n \in \mathbb{N}\} \cup \{0\}$. We show that E has the Heine–Borel property. Let $\mathcal{C} = \{U_\alpha : \alpha \in \mathcal{A}\}$ be an open cover for E. Because $0 \in E$ and $\mathcal{C}$ is an open cover of E, there exists $\alpha_0 \in \mathcal{A}$ such that $0 \in U_{\alpha_0}$. Because U_{α_0} is open, there exists $\varepsilon > 0$ such that $(-\varepsilon, \varepsilon) \subseteq U_{\alpha_0}$. Take $N \in \mathbb{N}$ such that $\frac{1}{N} < \varepsilon$. Then for all $n \ge N$, $\frac{1}{n} < \varepsilon$ and so, for all $n \ge N$, $\frac{1}{n} \in U_{\alpha_0}$. We have now found a single element of $\mathcal{C}$ containing all but finitely many points of E. Finding elements of $\mathcal{C}$ that contain the remaining points of E is easy. Because $\mathcal{C}$ is a cover for E, for each n with $1 \le n \le N-1$, there exists $\alpha_n \in \mathcal{A}$ such that $\frac{1}{n} \in U_{\alpha_n}$. Then $\{U_{\alpha_j} : 0 \le j \le N-1\}$ is a finite subcover of E.

Example 11.9. For a non-example, we show that $E = (0, \infty)$ does not have the Heine–Borel property. We need only exhibit an open cover of E without a finite subcover. Let $\mathcal{C} = \{(n-1, n+1) : n \in \mathbb{N}\}$. Clearly $\bigcup_{n=1}^{\infty}(n-1, n+1) = (0, \infty)$ and hence $\mathcal{C}$ covers E. Consider any finite subcollection $\mathcal{C}'$ of $\mathcal{C}$. Let $N = \max\{n : (n-1, n+1) \in \mathcal{C}'\}$. Because $\mathcal{C}'$ is a finite set, N exists and is a natural number. Now consider $x = N+1$. Clearly $x \in E$ but x is not in any of the sets in $\mathcal{C}'$. Thus $\mathcal{C}'$ is not a finite subcover of E. We conclude that E does not have the Heine–Borel property.

Exercise 11.14. Let E be a finite subset of $\mathbb{R}$. Show that E has the Heine–Borel property.

Exercise 11.15. Show that $(0, 1)$ does not have the Heine–Borel property.

The main result of this subsection is that the Heine–Borel property is equivalent to compactness.

Theorem 11.4 (Heine–Borel Theorem). *$K \subseteq \mathbb{R}$ is compact if and only if K has the Heine–Borel property.*

Proof. We first prove that if K has the Heine–Borel property, then K is closed and bounded, hence compact.

Suppose K has the Heine–Borel property. We show first that K is closed. Let x be an accumulation point of K. We must show that $x \in K$. Suppose not. Set $U_n = [x-\frac{1}{n}, x+\frac{1}{n}]^c$. Each U_n is open and $\bigcup_{n=1}^{\infty} U_n = \mathbb{R}\backslash\{x\}$. Thus $\{U_n : n \in \mathbb{N}\}$ covers K. Because K has the Heine–Borel property, some finite subcollection covers K. Let N be the largest index of an element of the subcollection. Because $U_1 \subseteq U_2 \subseteq \ldots$, we conclude that all sets in the subcollection are contained in U_N, hence the subcollection has union U_N. But then U_N itself contains K and so $(x - \frac{1}{N}, x + \frac{1}{N})$ is an open interval about x containing no point of K. This contradicts the assumption that x is an accumulation point of K, so we conclude that K is closed.

Next, we continue to suppose that K has the Heine–Borel property and we show that K is bounded. Let $U_n = (-n, n)$. Each U_n is open, and $\bigcup_{n=1}^{\infty} U_n = \mathbb{R}$, so $\{U_n : n \in \mathbb{N}\}$ is certainly an open cover for K. By assumption, some finite subcollection also covers K. If N is the largest index of a set in this subcollection, then U_N itself contains K, and K is bounded. We have thus completed the first direction in the proof of the theorem.

We now show that, if K is compact, then every open cover of K has a finite subcover. Let $\mathcal{C} = \{U_\alpha : \alpha \in \mathcal{A}\}$ be any open cover. We do not know how many sets are in this collection. It may be an uncountable collection. It turns out that *it is always possible to replace an uncountable cover of a set with a countable subcover.* This result does not use compactness; it relies on the fact that the collection of open intervals with rational endpoints is a countable set, the fact that we can use this countable set to select a countable subcollection from $\mathcal{C}$, and the fact that every x in K is in such an interval with rational endpoints. We leave the details to the next exercise.

As a consequence of Exercise 11.16, we may assume that $\mathcal{C}$ is a countable collection, that is, that $\mathcal{C} = \{U_n : n \in \mathbb{N}\}$. We must show that a finite subcollection of $\mathcal{C}$ covers K. Suppose not. Then for each n, K is not contained in $V_n = \bigcup_{j=1}^{n} U_j$. For each n, let $x_n \in K \setminus V_n$. Because $\{x_n\}$ is a sequence in K and K is compact, $\{x_n\}$ has a limit point x in K. This limit point is in at least one of the elements of $\mathcal{C}$. Let U_N be such a set. Then infinitely many terms of

$\{x_n\}$ are in U_N. But, by construction, none of $x_N, x_{N+1}, x_{N+2}, \ldots$ are in U_N. Because we have reached a contradiction, we conclude that a finite subcollection of $\mathcal{C}$ covers K. Thus K has the Heine–Borel property. □

Exercise 11.16. (Challenging) Let $\mathcal{C} = \{U_\alpha : \alpha \in \mathcal{A}\}$ be an arbitrary open cover for a set E with E not necessarily compact. The point of this exercise is to prove that there exists a countable collection $\mathcal{C}'$ of $\mathcal{C}$ covering E.

(a) Show that the set $\mathcal{I}$ of all open intervals with rational endpoints is a countable set.

(b) Let $\{I_n : n \in \mathbb{N}\}$ be an enumeration of $\mathcal{I}$. For each n, let U_n be an element of $\mathcal{C}$ containing I_n, if such a set exists. The collection of U_n thus obtained is countable. Show that it covers E.

11.3 A First Glimpse at the Notion of Measure

The previous sections explored important *topological* properties of the real line. Topology has to do with the structure of sets–whether all points in a set are completely surrounded by other points in the set, whether sequences in the set must have limit points in the set, and so on. Topological properties tend to be those that interact in predictable ways with continuous functions, as we will see in the next chapter.

In this section, we take a first look at *measure theoretic* properties of the real line. In measure theory, we are not so much concerned with how points in a set are arranged but rather with the size of the set. You might initially think that we have already discussed this topic, for in Chapter 7 we talked about cardinality as a way to say how big a set is. The problem is that if we use cardinality to measure subsets of the real line, we only have three answers: the set is finite, the set is countably infinite, and the set is uncountable. Most of the subsets we care about in analysis are intervals, and all intervals are uncountable and, in fact, in bijective correspondence with the entire real line. Thus we want a more sensitive measure of the size of a set that lets us compare intervals and more complicated sets.

Measure theory is a rather deep and difficult subject. Our aim is to give you just a tiny taste of it. Most notions in abstract mathematics are generalizations of simple or familiar notions. Measure is no exception. It is a generalization of the familiar notion of the length of an interval. We already have a good intuitive understanding of intervals, so we begin there. Then we use the notion of measure zero to explore some interesting "small" subsets of the real line.

11.3.1 Measuring intervals

We will use the term *interval* to refer to any of the sets (a, b), $[a, b]$, $(a, b]$, or $[a, b)$, where a and b are real numbers with $a < b$ or where $a = -\infty$ or $b = \infty$. In any of these cases, we call a and b the endpoints of the interval. We already have a natural way to measure intervals; we can talk about their lengths.

Definition 11.9. Let I be an interval with endpoints a and b. We define the **measure** of I to be the length of I, which is $b-a$. We denote the measure of I by $m(I)$.

Because we are allowing a or b or both to be infinite, we must make sense of expressions like $\infty - a$. We agree that, for any real number c, $\infty + c = \infty$. Also, we agree that $\infty + \infty = \infty$ and $-(-\infty) = \infty$. With these conventions established, we find, for example,

$$\begin{aligned} m((1,3)) &= 3-1=2 \\ m([1,\infty)) &= \infty - 1 = \infty \\ m((-\infty,4)) &= 4-(-\infty) = 4+\infty = \infty. \end{aligned}$$

These values are exactly what we expect; we think of the first interval as 2 units long, and we think of the second and third as having infinite length because they are unbounded.

Our first proposition summarizes the most basic properties of the length function for intervals. It is important to us because it gives the properties we want for a measure defined on more general sets.

Proposition 11.10. *Let I be an interval.*

(a) $m(I)$ is a non-negative real number or $m(I) = \infty$.

(b) (monotonicity) If J is also an interval and $I \subseteq J$, then $m(I) \leq m(J)$.

(c) (finite additivity) If $I = \bigcup_{n=1}^{N} I_n$ with the I_n disjoint intervals, then $m(I) = \sum_{n=1}^{N} m(I_n)$.

Proof. Part (a) follows immediately from the definition. For part (*b*), suppose that I has endpoints a and b and that J has endpoints c and d. Because $I \subseteq J$, $c \leq a$ and $b \leq d$. Then

$$m(I) = b-a \leq d-c = m(J),$$

as desired.

Finally consider part (c). We prove the statement for $N=2$. The general statement follows by induction and is left as an exercise. Thus suppose I is an interval and $I = I_1 \cup I_2$ with I_1 and I_2 disjoint intervals. We treat first the case in which $m(I)$ is finite. Suppose a and b are the left and right endpoints, respectively, of I and that a_j and b_j are the left and right endpoints of I_j. We may assume without loss of generality that $a_1 < a_2$. Then every point of I_1 is strictly less than every point of I_2 because the intervals are disjoint. Also, $a = a_1$, $b_1 = a_2$, and $b_2 = b$ because I is the union of I_1 and I_2. Thus

$$\begin{aligned} m(I) &= b-a \\ &= (b-a_2)+(a_2-a) \\ &= (b_2-a_2)+(b_1-a_1) \\ &= m(I_1)+m(I_2). \end{aligned}$$

We have proved the result when $m(I)$ is finite. When $m(I)$ is infinite, at least one of the endpoints of I is infinite, and thus at least one of the endpoints of I_1 or I_2 is infinite as well. In that case, at least one of $m(I_1)$ or $m(I_2)$ is infinite, and again $m(I) = m(I_1) + m(I_2)$. □

Exercise 11.17. Complete the proof of part (c) of Proposition 11.10 using induction of N. Be careful that when you remove an interval from $\bigcup_{n=1}^{N+1} I_n$ in order to invoke the induction hypothesis, the union you are left with *is an interval.*

In part (c) of Proposition 11.10, we considered the relationship between the measure of an interval I and the measures of disjoint intervals whose union is I. Because the sets we eventually hope to measure will not be disjoint unions of intervals, we seek to do something much more general. We seek a relationship between the measure of an interval I and the sum of the measures of the intervals in a *cover* of I.

Proposition 11.11 (finite subadditivity). *Let I be an interval and suppose that $I \subseteq \bigcup_{n=1}^{N} I_n$, where the I_n are intervals as well. Then $m(I) \leq \sum_{n=1}^{N} m(I_n)$.*

Proof. This proof is not hard, but it is rather tedious if we include all the details. We therefore merely sketch the argument. The idea is to replace each interval I_n with a set J_n that is either a subinterval of I_n or the empty set in such a way that the new collection $\{J_n\}$ is disjoint and has union I. (Note that although the empty set is not an interval, it is obvious that we should define its measure to be 0.) Part (c) of Proposition 11.10 will give $m(I) = \sum_{n=1}^{N} m(J_n)$. Then $J_n \subseteq I_n$ and monotonicity will give $\sum_{n=1}^{N} m(J_n) \leq \sum_{n=1}^{N} m(I_n)$.

How do we define the sets J_n? To begin with, if there are intervals I_n that do not intersect I, we may take the corresponding J_n to be the empty set and we will still have a collection whose union contains I but with a smaller sum of lengths. Thus it suffices to suppose for the remainder of the proof that each I_n intersects I. Suppose a and b are the left and right endpoints of I and that a_n and b_n are the left and right endpoints of I_n. We may assume by reordering if necessary that $a_1 \leq a_2 \leq \ldots \leq a_N$. Now, we let J_1 be the largest subinterval of I_1 that is contained in I. Proceed inductively. At step n, if I_n is contained in $\bigcup_{j=1}^{n-1} J_j$, set $J_n = \emptyset$. Otherwise, take J_n to be the largest subinterval of I_n that is contained in I and is disjoint from $J_1, \ldots, J_{n-1}$. By construction the J_n are disjoint. Furthermore, because the original collection covered I, one can show that $\bigcup_{n=1}^{N} J_n = I$. □

We would like to establish some less trivial properties of the length function for intervals. Again, the importance lies in the fact that these are properties that we will want to hold more generally when we define a measure on more complicated sets. In particular, we would like to be able to extend the finite additivity and finite subadditivity results to infinite collections of sets. Of course we only expect the generalizations to apply to *countably infinite* collections of sets because we do not know how to make sense of $\sum_{\alpha \in \mathcal{A}} m(I_\alpha)$ if the index set $\mathcal{A}$ is uncountable.

Proposition 11.12 (countable additivity). *Let I be an interval and suppose $I = \bigcup_{n=1}^{\infty} I_n$ with the I_n disjoint intervals. Then $m(I) = \sum_{n=1}^{\infty} m(I_n)$.*

Proof. The result holds whether $m(I)$ is finite or infinite, but we will only give the proof for $m(I)$ finite.

We prove the equality by proving two inequalities. First, fix N and consider $\bigcup_{n=1}^{N} I_n$. These intervals are disjoint, but their union is no longer I. Consider $I \setminus \bigcup_{n=1}^{N} I_n$. It is possible to write this set as a union of K disjoint intervals $J_1, \ldots, J_K$ with K at most $N+1$. Thus

$$I = \bigcup_{n=1}^{N} I_n \cup \bigcup_{k=1}^{K} J_k,$$

where the union is disjoint. By finite additivity,

$$\sum_{n=1}^{N} m(I_n) \leq \sum_{n=1}^{N} m(I_n) + \sum_{k=1}^{K} m(J_k) = m(I).$$

Thus the partial sums of the positive series $\sum_{n=1}^{\infty} m(I_n)$ are bounded above by $m(I)$ and the series converges, with

$$\sum_{n=1}^{\infty} m(I_n) \leq m(I).$$

The reverse inequality is harder to get. Let $\varepsilon > 0$ be given. As above, let a and b denote the endpoints of I and let a_n and b_n denote the endpoints of I_n. Because $m(I)$ is finite, both endpoints a and b of I are finite. The interval I may or may not be compact, but the slightly smaller interval $J = \left[a + \frac{\varepsilon}{2}, b - \frac{\varepsilon}{2}\right]$ is, and $m(J) = m(I) - \varepsilon$. Also, the intervals I_n may or may not be open, but the slightly larger intervals $J_n = \left(a - \frac{\varepsilon}{2^{n+1}}, b + \frac{\varepsilon}{2^{n+1}}\right)$ are. Note that $m(J_n) = m(I_n) + \frac{\varepsilon}{2^n}$.

Here's the important point: The collection $\{J_n\}$ is now an open cover for the compact set J. By the Heine–Borel Theorem (Theorem 11.4), some finite subcollection $\{J_1, \ldots, J_N\}$ covers J. By finite subadditivity,

$$m(J) \leq \sum_{n=1}^{N} m(J_n).$$

Thus

$$\begin{aligned} m(I) &= \varepsilon + m(J) \\ &\leq \varepsilon + \sum_{n=1}^{N} m(J_n) \\ &= \varepsilon + \sum_{n=1}^{N} \left(m(I_n) + \frac{\varepsilon}{2^n} \right) \\ &< 2\varepsilon + \sum_{n=1}^{N} m(I_n) \\ &\leq 2\varepsilon + \sum_{n=1}^{\infty} m(I_n). \end{aligned}$$

Because this inequality holds for all ε, we conclude that $m(I) \leq \sum_{n=1}^{\infty} m(I_n)$. Our two inequalities prove the proposition. □

Proposition 11.13 (countable subadditivity). *Let I be an interval and suppose $I \subseteq \bigcup_{n=1}^{\infty} I_n$. Then $m(I) \leq \sum_{n=1}^{\infty} m(I_n)$.*

Proof. We may approach this proof in the same way we approached the proof of finite subadditivity. We omit the details. □

Although some of these proofs are technical and tedious, when we stand back and look again at the statements of the propositions, we see that they say something very simple. The length function for intervals behaves just the way our intuition suggests it should. An interval is longer than any of its subintervals. If we break an interval into pieces, the lengths of the pieces add up to the length of the interval. If an interval is covered by a collection of intervals, the length of the interval is less than the sum of the lengths of the intervals in the cover.

If these propositions about the length function for intervals are so intuitive, why do we bother to state and prove them? Remember, we are looking for a way to measure more complicated subsets of $\mathbb{R}$. Our work with intervals helps us in three ways. First, because any class of sets we measure will certainly include the intervals, we can get a feeling for what properties a general notion of measure might satisfy. Second, by proving these propositions, we may start to get some insight into proof techniques that might work more generally. Third, perhaps we can actually use intervals in our definition of measures of more general sets. If such an approach is possible, propositions about measure in general will follow from the corresponding propositions for intervals.

11.3.2 Measure zero

How would we measure a more complicated subset E of $\mathbb{R}$? Here's the idea: Cover E with a countable collection $\{I_n : n \in \mathbb{N}\}$ of intervals. If measure is to satisfy monotonicity in general, then $\sum_{n=1}^{\infty} m(I_n)$ should be an upper bound for $m(E)$. If we take the infimum over all possible coverings of E by intervals, this number is a good candidate for $m(E)$. There is a problem. If we allow ourselves to measure all sets in this way, we won't get a measure with all the properties we desire. Thus we must restrict the class of sets we measure. We will not discuss here how we restrict the class of sets. We only comment that it can be done in such a way to still include a large number of sets. In particular, the class includes all open and closed sets and all sets obtained by countably many set operations on open and closed sets.

One class of measurable sets is easy to describe and fun to explore, so we do so here.

Definition 11.10. A subset E of $\mathbb{R}$ has **measure zero** if, for every $\varepsilon > 0$, there exists a collection $\{I_n : n \in \mathbb{N}\}$ of intervals covering E with $\sum_{n=1}^{\infty} m(I_n) < \varepsilon$.

We consider a number of examples. Of course, we think of sets of measure zero as small. One of the purposes of all the examples is to show the diversity of sets that have measure zero.

Example 11.10. Let E be a finite set, so that $E = \{x_1, x_2, \ldots, x_N\}$, where each x_n is a real number. We claim that $m(E) = 0$. Let $\varepsilon > 0$ be given and set $I_n = (x_n - \frac{\varepsilon}{4N}, x_n + \frac{\varepsilon}{4N})$. These intervals cover E, and

$$\sum_{n=1}^{N} m(I_n) = \sum_{n=1}^{N} \frac{\varepsilon}{2N} = \frac{\varepsilon}{2} < \varepsilon.$$

Therefore, by definition, $m(E) = 0$.

Example 11.11. For our next example, we consider a much larger set, namely the set $\mathbb{Q}$ of rational numbers. We currently know several things about the size of this set. In Chapter 7, we proved that $\mathbb{Q}$ is countable. Because $\mathbb{R}$ is uncountable, this result indicates a sense in which $\mathbb{Q}$ is small. In Chapter 9, we proved in Corollary 9.1 that $\mathbb{Q}$ is dense in $\mathbb{R}$, that is, that every open interval about every real number contains at least one rational number. This result indicates a sense in which $\mathbb{Q}$ is large. Now we may consider the measure of $\mathbb{Q}$. We will show that $m(\mathbb{Q}) = 0$. Because $m(\mathbb{R}) = \infty$, this result indicates that "almost every" real number is irrational.

Let $\varepsilon > 0$ be given. Because $\mathbb{Q}$ is countable, we can find an enumeration $\{r_1, r_2, r_3, \ldots\}$ of $\mathbb{Q}$. For each n, we will define a small open interval centered at r_n, and so $\mathbb{Q}$ will certainly be covered by the collection of all such intervals. We want the lengths of the intervals to add up to less than ε. If we took all intervals to be the same length, because there are infinitely many of them, no matter how small we make them, their lengths would sum to ∞. Our only hope is to make each interval smaller than the previous in such a way that the sum of the lengths remains less than ε. Thus we let $I_n = (r_n - \frac{\varepsilon}{2^{n+2}}, r_n + \frac{\varepsilon}{2^{n+2}})$, so that $m(I_n) = \frac{\varepsilon}{2^{n+1}}$. Then $\mathbb{Q} \subseteq \bigcup_{n=1}^{\infty} I_n$ and

$$\sum_{n=1}^{\infty} m(I_n) = \sum_{n=1}^{\infty} \frac{\varepsilon}{2^{n+1}} = \frac{\frac{\varepsilon}{4}}{1 - \frac{1}{2}} = \frac{\varepsilon}{2} < \varepsilon.$$

Thus according to our definition, $m(\mathbb{Q}) = 0$.

Exercise 11.18. Explain why the argument in Example 11.11 proves that every countable set has measure zero.

11.3.3 The Cantor set

Are there any uncountable sets with measure zero? So far we have only met two kinds of uncountable subsets of $\mathbb{R}$, namely intervals and complements of countable sets. In this subsection we discuss an unusual set, the Cantor middle-thirds set, with many beautiful properties. You may have already met this set if you did Problem 4 in the previous chapter. If you have not, you should do that problem now.

The Cantor set is the result of an infinite process. The starting point is the set $C_0 = [0,1]$. For each n, C_n is a union of a finite number of closed intervals. For $n \geq 1$, we obtain C_n from C_{n-1} by removing from each closed interval comprising C_{n-1} the open interval that is its middle third. Thus at step 1 we remove the interval $\left(\frac{1}{3}, \frac{2}{3}\right)$ from C_0 to obtain $C_1 = \left[0, \frac{1}{3}\right] \cup \left[\frac{2}{3}, 1\right]$. At step 2, we remove the intervals $\left(\frac{1}{9}, \frac{2}{9}\right)$ and $\left(\frac{7}{9}, \frac{8}{9}\right)$ from the intervals in C_1 to obtain $C_2 = \left[0, \frac{1}{9}\right] \cup \left[\frac{2}{9}, \frac{1}{3}\right] \cup \left[\frac{2}{3}, \frac{7}{9}\right] \cup \left[\frac{8}{9}, 1\right]$, and so on. We define the Cantor set to be $C = \bigcap_{n=0}^{\infty} C_n$.

Each C_n is a closed set because it is a finite union of closed intervals. The Cantor set itself is an intersection of a countable collection of closed sets. We do not have a general result about countable intersections of closed sets, but we can determine the topological properties of the Cantor set.

Proposition 11.14. *The Cantor set C is compact.*

Proof. We will prove that C is compact by proving that it is closed and bounded. (See Theorem 11.3.) Boundedness is obvious because $C \subseteq [0,1]$.

To prove that C is closed, we prove that its complement is open. By De Morgan's laws (Proposition 4.1), $C^c = \bigcup_{n=0}^{\infty} C_n^c$. Thus C^c is the union of $(-\infty, 0)$, $(1, \infty)$, and all the open intervals removed during all the steps of the construction. Because an arbitrary union of open sets is open by Proposition 11.2, C^c is indeed open. □

How big is the Cantor set? We could answer this question in several ways, depending on what we mean by size. We could talk about the cardinality of the set, or we could talk about measure. We do both.

Proposition 11.15. *The Cantor set C is uncountable.*

Proof. We must construct a bijection between C and a set we know to be uncountable. For the latter, we take the set S of all infinite sequences of zeros and ones. Let $x \in C$. Then $x \in C_n$ for all n. We use this fact to define a sequence s indexed by the natural numbers as follows: Suppose $n \geq 1$. Because $x \in C_{n-1}$, x is in one of the closed intervals I_k making up C_{n-1}. At step n, the middle third of I_k is removed, leaving a left subinterval L_k and a right subinterval R_k

of I_k. Because the point x is not removed, x is in one of these two subintervals. Let $s_n = 0$ if $x \in L_k$ and let $s_n = 1$ if $x \in R_k$. The sequence s thus obtained essentially gives an address for the point x of C. We now have a function from C to S, and it is not hard to see that this function is a bijection. Thus C is uncountable. $\square$

Proposition 11.16. *The Cantor set C has measure zero.*

Proof. Let $\varepsilon > 0$ be given. For all n, $C \subseteq C_n$. Furthermore, each C_n is a union $\bigcup_{1 \leq k \leq 2^n} I_{n,k}$ of 2^n closed intervals, each of length 3^{-n}. In other words, $\{I_{n,k} : 1 \leq k \leq 2^n\}$ is a cover for C. Take $N \in \mathbb{N}$ such that $\left(\frac{2}{3}\right)^N < \varepsilon$. Then $\{I_{N,k} : 1 \leq k \leq N\}$ is the desired cover of C with $\sum_{k=1}^{2^N} m(I_{N,k}) = \left(\frac{2}{3}\right)^N < \varepsilon$. $\square$

Taken together, the previous propositions are remarkable; we have a set whose cardinality is that of the real numbers but which is so small that its measure is zero! The Cantor set is one of my favorite examples from elementary analysis because of these seemingly paradoxical properties. It is the example that first made me fall in love with the subject of analysis. The exercises and problems let you continue your exploration of this beautiful mathematical object.

Exercise 11.19. Show that every point of C is an accumulation point of C. (Given $x \in C$, you might look for some sequences in C that converge to x and argue that at least one of them has infinitely many distinct terms.) A set in which every point is an accumulation point is called a *perfect* set.

11.4 Problems

1. For each of the following, either give an example of such a set or explain why no such set exists.

   (a) An infinite collection $\{U_j : j \in \mathbb{N}\}$ of open sets whose intersection is open.

   (b) A closed set with no accumulation points.

   (c) An open set with no accumulation points.

   (d) A closed set whose complement is also closed.

   (e) A non-empty subset of $[0, 1]$ with no accumulation points.

   (f) An infinite subset of $[0, 1]$ with no accumulation points.

2. In this problem, we explore the topological properties of the set $\mathbb{Q}$ of rational numbers.

   (a) Let x be a rational number. Show that, for every $\varepsilon > 0$, $(x - \varepsilon, x + \varepsilon)$ contains a rational number different from x.

(b) Let x be a rational number. Show that, for every $\varepsilon > 0$, $(x-\varepsilon, x+\varepsilon)$ contains an irrational number.

(c) Now let x be an irrational number. Show that, for every $\varepsilon > 0$, $(x-\varepsilon, x+\varepsilon)$ contains both a rational number and an irrational number different from x.

(d) Is $\mathbb{Q}$ open? Closed?

(e) Which elements of $\mathbb{R}$ are accumulation points of $\mathbb{Q}$?

3. For each of the following, give an example of a sequence with the stated property or explain why no such sequence exists.

   (a) A sequence with four limit points.

   (b) A divergent sequence with only one limit point.

   (c) A convergent sequence with two distinct limit points.

4. For each of the following, either give an example of a subset of $\mathbb{R}$ with the stated property or explain why no such set exists.

   (a) A closed set that is not compact.

   (b) A compact set that is not closed.

   (c) A bounded set that is not compact.

   (d) A compact set with no accumulation points.

5. (Challenging) For each of the following, either give an example of a sequence with the stated property, or explain why no such sequence exists.

   (a) A sequence that has every rational number as a limit point. (These needn't be the only limit points.)

   (b) A sequence that has every real number as a limit point.

6. Let $E = \{\frac{(-1)^n}{n} : n \in \mathbb{N}\}$. Does E have the Heine–Borel property?

7. For each of the following, either give an example of a set with the stated property or explain why no such set exists.

   (a) A countably infinite set with the Heine–Borel property.

   (b) A countably infinite set that does not have the Heine–Borel property.

   (c) An open cover of $(-1, 1)$ with no finite subcover.

   (d) A non-compact set that has an open cover with a finite subcover.

8. For each of the following, either give an example of a set with the stated property or explain why no such set exists.

   (a) A non-empty open set with measure zero.

   (b) A non-empty closed set with measure zero.

9. What can you say about the base 3 representations of the points in the Cantor set?

Programming Project. Consider the sequence $\{\sin(n)\}$.

1. Plot the first 1000 terms and make a conjecture about its limit points.

2. Can you use the theorems of this chapter to prove that the sequence has a limit point? Either do so or explain why it is not possible.

3. Write a program that takes as input a suspected value of a limit point L and a value for ε and gives the following as output: (1) a list of all indices of terms of $\{\sin(n)\}$ within ε of L for $0 \leq n \leq 10000$, and (2) a plot of the subsequence thus extracted.

4. Use your program to further explore the sequence and modify your conjecture about the set of limit points if necessary.

5. Now consider the sequence $\{\sin\left(\frac{\pi n}{20}\right)\}$. Answer all of the questions above for this sequence.

Chapter 12

Continuous Functions

From calculus, you already have an intuitive understanding of a continuous function as a function whose graph has no holes, jumps, or asymptotes. The goals of this chapter are to make this notion precise and to establish the basic theorems about continuous functions.

12.1 Sequential Continuity

Consider $f\colon (c,d) \to \mathbb{R}$. We first define continuity of f at a single point a of the domain. In order that f not have a "jump" or "hole" at a, f should have the property that inputs to the function near a produce outputs of the function near $f(a)$. Here is the precise definition.

Definition 12.1. Let $D \subseteq \mathbb{R}$, let $f\colon D \to \mathbb{R}$, and let $a \in D$. Then f is **continuous at** a if, for every sequence $\{x_n\}$ in D with $\lim_{n\to\infty} x_n = a$, $\lim_{n\to\infty} f(x_n) = f(a)$.

Thus continuous functions send convergent sequences to convergent sequences. Sometimes the notion of continuity presented in Definition 12.1 is called *sequential continuity* because of the role played by sequences in the definition. We do not make this distinction, though we do show later that this definition is equivalent to another notion that could have been taken as the definition of continuity at a point.

12.1.1 Exploring sequential continuity graphically and numerically

As always, we illustrate the definition with some examples. Consider two functions f and g on $\mathbb{R}$ given by

$$f(x) = \begin{cases} \frac{x^3-4x}{x-2} & x \neq 2 \\ 8 & x = 2 \end{cases} \tag{12.1}$$

Figure 12.1: First 50 terms of the sequence $\{y_n\}$.

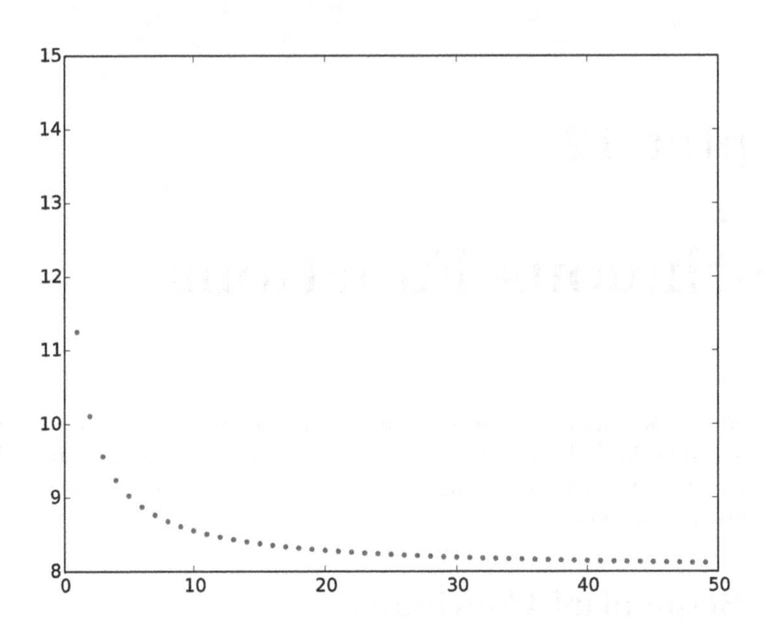

and

$$g(x) = \begin{cases} \frac{x^3-4x}{x-2} & x \neq 2 \\ 7 & x = 2 \end{cases}. \tag{12.2}$$

These functions agree for every input except 2; $f(2) = 8$, whereas $g(2) = 7$. We consider the continuity of each of these functions at $a = 2$. Our definition says we need to consider sequences $\{x_n\}$ with limit 2 and determine whether or not the sequence of function values has a limit equal to the value of the function at 2.

Consider, for example, the sequence $\{x_n\}$ for which $x_n = 2 + \frac{1}{n}$. Because 2 is never a term of the sequence and because the functions f and g agree at all points other than 2, the sequences $\{f(x_n)\}$ and $\{g(x_n)\}$ are identical. For convenience, call this common sequence $\{y_n\}$. Does $\{y_n\}$ converge? If so, to what? We have considered questions like this one many times, most notably in Chapter 8. Of course, we can explore the convergence of this sequence graphically or numerically. Figure 12.1 shows a graph of the first 50 terms of the sequence $\{y_n\}$. If we generate many terms of the sequence, we find that y_{998}, y_{999}, and y_{1000} are

$$8.00601303, 8.00600701, 8.00600100.$$

The evidence leads us to conjecture that $\lim_{n\to\infty} y_n = 8$. If this conjecture is correct, what does it tell us about the continuity of f and g at 2? We would be able to conclude that g is *not* continuous at 2 because we would have an example of a sequence $\{x_n\}$ converging to 2 with $\{g(x_n)\}$ converging to a number different from $g(2)$, which is 7. However, we would not be able to make a conclusion about the continuity of f. The definition requires us to show that for *all* sequences $\{x_n\}$ approaching 2, $\{f(x_n)\}$ approaches $f(2)$. We have only considered a single sequence here.

Exercise 12.1. Substitute the expression $x_n = 2 + \frac{1}{n}$ for x in the equation $f(x) = \frac{x^3-4x}{x-2}$, simplify, and then take the limit to show that $\lim_{n\to\infty} f(x_n) = 8$.

12.1.2 Proving that a function is continuous

Although graphs and numerical calculations help us explore functions to make conjectures about their continuity, they do not give us proofs of those conjectures. Proofs come from using the definition of continuity and what we know about sequences. In this subsection, we give several proofs of the continuity of specific functions.

Example 12.1. Consider $f\colon \mathbb{R} \to \mathbb{R}$ defined by $f(x) = x^2 + 1$. Our experience from calculus suggests that this function is continuous at every real number a. To prove this claim, we must show that, if $\{x_n\}$ has limit a, then $\{f(x_n)\}$ has limit $f(a)$. Note that $f(x_n) = x_n^2 + 1$ and $f(a) = a^2 + 1$. Thus we must show that if $\lim_{n\to\infty} x_n = a$, then $\lim_{n\to\infty}(x_n^2 + 1) = a^2 + 1$.

Because $\{x_n\}$ has limit a, by the limit laws (Proposition 8.3), the sequence with n-th term $x_n \cdot x_n = x_n^2$ has limit $a \cdot a = a^2$. Then, again by the limit laws and the fact that the constant sequence with n-th term 1 has limit 1,

$$\begin{aligned}\lim_{n\to\infty}(x_n^2 + 1) &= \lim_{n\to\infty} x_n^2 + \lim_{n\to\infty} 1\\ &= a^2 + 1.\end{aligned}$$

The conclusion follows.

Exercise 12.2. Show that $f(x) = 3x - 2$ is continuous at any $a \in \mathbb{R}$.

Exercise 12.3. Part (c) of Proposition 8.3 says that if a and b are convergent sequences, then the limit of the product sequence is the product of the limits. We want to generalize this proposition to a product of k convergent sequences. Use induction on k to prove that if $b^{(j)} = \{b_n^{(j)}\}$, $1 \le j \le k$, are convergent sequences with limits L_j, $1 \le j \le k$, then the product sequence with n-th term $b_n^{(1)} b_n^{(2)} \dots b_n^{(k)}$ is convergent with limit $L_1 L_2 \cdots L_k$.

Exercise 12.4. Use the previous exercise to show that, for any $n \in \mathbb{N}$, $f(x) = x^n$ is continuous at every $a \in \mathbb{R}$.

We consider a second example. As with the previous example and exercises, limit laws feature prominently.

Example 12.2. Consider $f\colon (0, \infty) \to \mathbb{R}$ defined by $f(x) = \frac{1}{x}$. We claim that if $a \in (0, \infty)$, f is continuous at a. Indeed, let $\{x_n\}$ be any sequence in $(0, \infty)$

such that $\lim_{n\to\infty} x_n = a$. Because each $x_n \neq 0$, $\frac{1}{x_n}$ is defined for all n. Because $\lim_{n\to\infty} x_n \neq 0$, by the limit laws,

$$\begin{aligned} \lim_{n\to\infty} f(x_n) &= \lim_{n\to\infty} \frac{1}{x_n} \\ &= \frac{1}{a} \\ &= f(a). \end{aligned}$$

Thus f is indeed continuous at a.

Exercise 12.5. Consider $f\colon \mathbb{R} \to \mathbb{R}$ defined by $f(x) = \frac{x}{x^2+1}$. Show that f is continuous at every $a \in \mathbb{R}$.

12.1.3 Proving that a function is discontinuous

Often, proving that a function is discontinuous at a point a is easy because all we need to do is find a single sequence $\{x_n\}$ converging to a for which $\{f(x_n)\}$ does not converge to $f(a)$.

Example 12.3. Consider $f\colon \mathbb{R} \to \mathbb{R}$ defined by

$$f(x) = \begin{cases} 0 & x < 0 \\ 1 & x \geq 0 \end{cases}.$$

Our intuition is that this function is not continuous at 0 because it has a jump.

Because $f(0) = 1$, all we need to do is find a sequence $\{x_n\}$ converging to 0 for which $\{f(x_n)\}$ does not have limit 1. This is easy; take $x_n = -\frac{1}{n}$. Because $x_n < 0$ for all n, $f(x_n) = 0$ for all n. Thus the sequence $\{f(x_n)\}$ has limit 0, not 1. Thus f is discontinuous at 0.

Our intuition from calculus also tells us that if a function is unbounded on every interval containing a, it is discontinuous at a no matter how we define it there. Our next example illustrates this situation.

Example 12.4. Consider $f\colon \mathbb{R} \to \mathbb{R}$ defined by

$$f(x) = \begin{cases} \frac{1}{x^2} & x \neq 0 \\ b & x = 0 \end{cases}.$$

We claim there is no value for b for which f is continuous at 0. Indeed, consider the sequence $\{x_n\} = \{\frac{1}{n}\}$. This sequence clearly has limit 0, but $\{f(x_n)\} = \{n^2\}$ is unbounded, hence divergent.

Exercise 12.6. Consider $f\colon \mathbb{R} \to \mathbb{R}$ defined by

$$f(x) = \begin{cases} x^3 & x \leq 1 \\ 2x^2 & x > 1 \end{cases}.$$

Show that f is discontinuous at 1.

12.1.4 First results

In our previous subsections, the limit laws for sequences were an important tool in our proofs. It is therefore not surprising that the limit laws can be used to prove that certain combinations of continuous functions are continuous.

Proposition 12.1. *Let f and g be real-valued functions defined on some common domain $D \subseteq \mathbb{R}$ containing the point a and suppose that f and g are continuous at a. Then*

(a) for any real A, Af is continuous at a.

(b) $f + g$ is continuous at a.

(c) fg is continuous at a.

(d) if $g(a) \neq 0$, $\frac{f}{g}$ is continuous at a.

Proof. We give the proofs of parts (b) and (d) and leave the proofs of parts (a) and (c) to the reader.

Consider part (b). Let $\{x_n\}$ be any sequence in D with limit a. We must show that $\{(f+g)(x_n)\}$ has limit $(f+g)(a)$. Note that, by definition,

$$(f+g)(x_n) = f(x_n) + g(x_n)$$

and

$$(f+g)(a) = f(a) + g(a).$$

Because f and g are continuous at a, $\{f(x_n)\}$ is convergent with limit $f(a)$ and $\{g(x_n)\}$ is convergent with limit $g(a)$. Thus by the limit laws, $\{f(x_n) + g(x_n)\}$ is convergent with limit $f(a) + g(a)$. The continuity of $f + g$ at a follows.

Next consider part (d). Again, let $\{x_n\}$ be an arbitrary sequence in D with limit a. By the continuity of f and g at a, $\{f(x_n)\}$ converges to $f(a)$ and $\{g(x_n)\}$ converges to $g(a)$. Because $g(a) \neq 0$, there exists N such that, for all $n \geq N$, $g(x_n) \neq 0$. Thus for $n \geq N$, $\left(\frac{f}{g}\right)(x_n) = \frac{f(x_n)}{g(x_n)}$ makes sense. By the limit laws, $\left\{\frac{f(x_n)}{g(x_n)}\right\}$ is convergent with limit $\frac{f(a)}{g(a)} = \left(\frac{f}{g}\right)(a)$. Thus $\frac{f}{g}$ is continuous at a. □

Exercise 12.7. Prove parts (a) and (c) of Proposition 12.1.

Just as the limit laws for sequences allowed us to establish the convergence of a wider range of sequences without having to revert to the definition of convergence, Proposition 12.1 allows us to establish continuity properties for a large class of familiar functions. In particular, because polynomials and rational functions are formed from the basic continuous function $f(x) = x$ and $g(x) = c$ through a finite number of additions, subtractions, multiplications, and divisions, all such functions are continuous at every point of $\mathbb{R}$, with the exception of points at which a denominator is zero. We leave the simple details of these arguments to the exercises.

Exercise 12.8. Use induction to prove that if $f_j\colon D \to \mathbb{R}$, $1 \le j \le k$, are continuous at a, then $\sum_{j=1}^{k} f_j$ and $\prod_{j=1}^{k} f_j$ are continuous at a.

Exercise 12.9. Prove that if f is a polynomial, f is continuous at every $a \in \mathbb{R}$.

Exercise 12.10. Let $r = \frac{p}{q}$ be a rational function, so that p and q are polynomials. Note that r is not defined at a if $q(a) = 0$. Prove that r is continuous at every $a \in \mathbb{R}$ at which it is defined.

12.2 Related Notions

12.2.1 The ε-δ condition

We now prove a theorem that gives a condition equivalent to continuity of a function at a point.

Theorem 12.1. *Let $f\colon D \to \mathbb{R}$ and let $a \in D$. Then f is continuous at a if and only if, for every $\varepsilon > 0$, there exists $\delta > 0$ such that, for all $x \in D$ with $|x - a| < \delta$, $|f(x) - f(a)| < \varepsilon$.*

Proof. First we prove that if f is continuous at a, then the ε-δ condition is satisfied. We prove this assertion by proving the contrapositive. Thus we suppose that there exists $\varepsilon > 0$ such that, for all $\delta > 0$, there exists $x \in D$ with $|x - a| < \delta$ but $|f(x) - f(a)| \ge \varepsilon$. In particular, for this ε, for every $n \in \mathbb{N}$, there exists $x_n \in D$ with $|x_n - a| < \frac{1}{n}$ but $|f(x_n) - f(a)| \ge \varepsilon$. By construction, $\{x_n\}$ is a sequence with limit a, but $\{f(x_n)\}$ does not have limit $f(a)$. Thus f is not continuous at a.

Conversely, we suppose that the ε-δ definition is satisfied by f at a and we prove that f is continuous at a. Let $\{x_n\}$ be a sequence in D with limit a. Consider $\{f(x_n)\}$. We must show that this sequence has limit $f(a)$. Let $\varepsilon > 0$ be given. We must show that there exists N such that, for all $n \ge N$, $|f(x_n) - f(a)| < \varepsilon$.

Now, because the ε-δ condition is satisfied, for our given ε, there exists $\delta > 0$ such that, for every $x \in D$ with $|x - a| < \delta$, $|f(x) - f(a)| < \varepsilon$. Because $\{x_n\}$ has limit a, for this δ, there is an N such that, for all $n \ge N$, $|x_n - a| < \delta$. Thus for all $n \ge N$, $|f(x_n) - f(a)| < \varepsilon$, as desired. ☐

Many analysis books take the ε-δ condition to be the definition of continuity for f at a. Of course, because this condition is equivalent to sequential continuity at a, it does not matter which one we take as the definition and which one we then prove as the theorem. For most examples, we can obtain a proof of the continuity or discontinuity of the function using either condition, though which one we choose will affect the flavor of the argument. We consider an example.

Example 12.5. Even though we already know that polynomial functions are continuous, we will prove again that $f(x) = x^2 + 1$ is continuous at any a in order to illustrate the flavor of continuity arguments using the ε-δ condition.

Let $\varepsilon > 0$ be given. We must show that there exists $\delta > 0$ such that $|x^2 + 1 - (a^2 + 1)| < \varepsilon$ whenever $|x - a| < \delta$. Observe,

$$|x^2 + 1 - (a^2 + 1)| = |x^2 - a^2| = |x + a||x - a|.$$

If the last expression were bounded by $C|x - a|$ for some positive C, we could take $\delta = \frac{\varepsilon}{C}$. If we think of x as being allowed to range over all of $\mathbb{R}$, there is no such positive upper bound for $|x+a|$. But we do not need to think of x as ranging over all of $\mathbb{R}$; we need only show that *when* $|x - a| < \delta$, $|f(x) - f(a)| < \varepsilon$. In other words, we only need an upper bound for $|x+a|$ that holds when $|x-a| < \delta$. We still have a bit of a problem, for we have not yet chosen δ.

Here is the key idea: We need only show that *some* $\delta > 0$ exists for which $|x-a| < \delta$ implies $|f(x)-f(a)| < \varepsilon$. Thus we may always place some restriction on the size of the δ we will consider. Let's suppose we will only consider δ's that are less than 1. Then for any such δ, $|x - a| < \delta$ will imply $|x - a| < 1$, which, by the reverse triangle inequality, implies $|x| - |a| < 1$, or $|x| < |a| + 1$. Thus if $\delta < 1$, $|x - a| < \delta$ implies

$$\begin{aligned} |f(x) - f(a)| &= |x + a||x - a| \\ &\leq (|x| + |a|)|x - a| && \text{(by the triangle inequality)} \\ &\leq (2|a| + 1)|x - a| && \text{(because } |x| < |a| + 1\text{)}. \end{aligned}$$

This last expression is less than ε if $|x - a| < \frac{\varepsilon}{2|a|+1}$. We thus take $\delta = \min\left\{1, \frac{\varepsilon}{2|a|+1}\right\}$.

Exercise 12.11. (a) For $|x-a| < 1$, find an upper bound for $|(x^2+a^2)(x+a)|$ that depends on a but not on x.

(b) Prove that $f(x) = x^4$ is continuous at any $a \in \mathbb{R}$ by verifying the ε-δ condition.

For us, taking Definition 12.1 as the definition of continuity of f at a made it easy to prove that polynomials are continuous because we had already done all the hard work necessary to establish the limit laws for sequences. For functions that are not polynomials or rational functions, we do not have limit laws to fall back on, and we may find it easier to establish continuity properties of a function by using the ε-δ condition. We illustrate with an example.

Example 12.6. Consider $f\colon (0, \infty) \to \mathbb{R}$ defined by $f(x) = \frac{1}{\sqrt{x}}$. We claim that f is continuous at any $a \in (0, \infty)$. Let $\varepsilon > 0$ be given. We must show that there exists $\delta > 0$ such that $\left|\frac{1}{\sqrt{x}} - \frac{1}{\sqrt{a}}\right| < \varepsilon$ whenever $x \in (0, \infty)$ and $|x - a| < \delta$.

Observe,

$$\begin{aligned}\left|\frac{1}{\sqrt{x}}-\frac{1}{\sqrt{a}}\right| &= \left|\frac{\sqrt{a}-\sqrt{x}}{\sqrt{x}\sqrt{a}}\right| \\ &= \left|\frac{(\sqrt{a}-\sqrt{x})(\sqrt{a}+\sqrt{x})}{\sqrt{x}\sqrt{a}(\sqrt{a}+\sqrt{x})}\right| \\ &= \frac{1}{\sqrt{a}\sqrt{x}(\sqrt{a}+\sqrt{x})}\cdot|x-a| \\ &< \frac{1}{\sqrt{a}\sqrt{x}(\sqrt{a}+\sqrt{0})}|x-a| \\ &= \frac{1}{a\sqrt{x}}|x-a|.\end{aligned}$$

We make a preliminary restriction on δ that we may use to estimate the factor $\frac{1}{a\sqrt{x}}$ for x near a. In the example above, we simply required δ to be less than 1, and this restriction was enough to get an upper bound on the factor multiplying $|x-a|$ for $|x-a|<\delta$. In the current example, however, in order to get an upper bound on the factor $\frac{1}{a\sqrt{x}}$, we need a *lower bound* on x so that x can not be too close to 0. Suppose we require $\delta < \frac{a}{2}$. Then $|x-a|<\delta$ implies, by the reverse triangle inequality,

$$|a|-|x|<\frac{a}{2} \iff |x|>|a|-\frac{a}{2}=\frac{a}{2}$$

because a is positive. With this restriction on δ, $x \in (0,\infty)$ and $|x-a|<\delta$ imply

$$\begin{aligned}|f(x)-f(a)| &< \frac{1}{a\sqrt{x}}|x-a| \\ &< \frac{1}{a\sqrt{\frac{a}{2}}}|x-a| \\ &= \frac{\sqrt{2}}{a\sqrt{a}}|x-a|.\end{aligned}$$

This last expression is less than ε if $|x-a|<\frac{\varepsilon a\sqrt{a}}{\sqrt{2}}$. Thus take $\delta=\min\left\{\frac{a}{2},\frac{\varepsilon a\sqrt{a}}{\sqrt{2}}\right\}$.

Exercise 12.12. Use the ε-δ condition to prove that $f(x)=\sqrt{x}$ is continuous at any $a \in [0,\infty)$.

12.2.2 Uniform continuity

So far in this chapter we have discussed continuity of a function *at a point.* Often we want to step back and consider the behavior of a function on a set.

Definition 12.2. If $f\colon D \to \mathbb{R}$, we say that f is **continuous on** D if f is continuous at every point of D. In terms of the ε-δ condition, f is continuous on D if, for every $a \in D$, for every $\varepsilon>0$, there exists $\delta>0$ such that, for every $x \in D$ with $|x-a|<\delta$, $|f(x)-f(a)|<\varepsilon$.

Figure 12.2: A function that is not uniformly continuous on $(0,1)$.

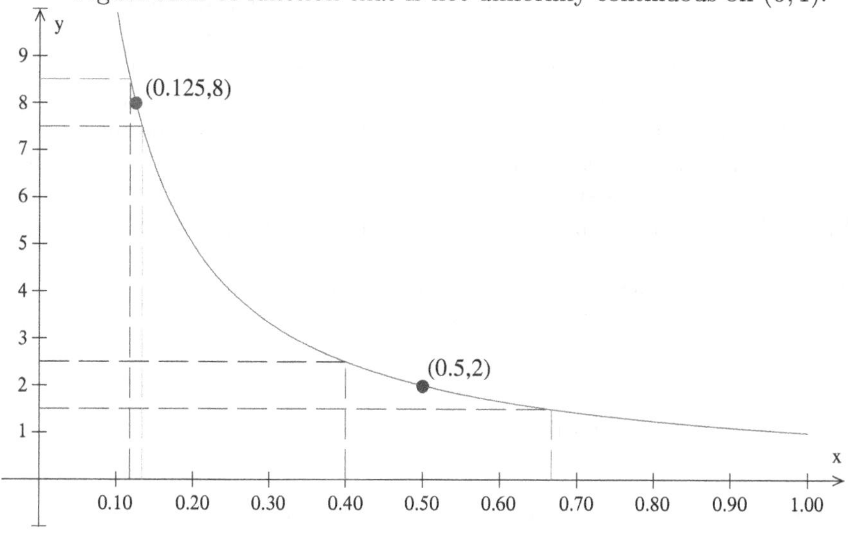

We have shown, for example, that every polynomial is continuous on $\mathbb{R}$ and that every rational function is continuous on its natural domain of definition.

In this subsection, we define the notion of *uniform continuity*, which is a property of a function *and a set*. Compare the next definition with Definition 12.2.

Definition 12.3. Let $f\colon D \to \mathbb{R}$. Then f is **uniformly continuous on** D if, for every $\varepsilon > 0$, there exists $\delta > 0$ such that, for every $a, x \in D$ with $|x-a| < \delta$, $|f(x) - f(a)| < \varepsilon$.

It is immediate that if f is uniformly continuous on D, it is continuous on D, for the ε-δ condition holds at every a in D. Uniform continuity is a stronger condition, however, for it requires that for a given $\varepsilon > 0$, there exists a *single* $\delta > 0$ with the property that *for every* $a, x \in D$ with $|x - a| < \delta$, $|f(x) - f(a)| < \varepsilon$. In other words, δ does not depend on a, as it appears to have above. Of course, in order to see that uniform continuity on D is in fact stronger than continuity on D, we must give an example of a function that is continuous but not uniformly continuous on some set D.

Example 12.7. We know that $f\colon (0,1) \to \mathbb{R}$ defined by $f(x) = \frac{1}{x}$ is continuous on $(0,1)$. We claim that f is not uniformly continuous on $(0,1)$.

Before proving the claim, let's try to understand conceptually why it is true. The definition of uniform continuity says that, given an $\varepsilon > 0$, we can find a δ independent of a such that, as long as x is within δ of a, $f(x)$ is within ε of $f(a)$. Let's look at a graph of $f(x) = \frac{1}{x}$. (See Figure 12.2.) We fix $\varepsilon = \frac{1}{2}$ and see, for different values of a, how close x must be to a in order for $f(x)$ to be within $\frac{1}{2}$ of $f(a)$. First consider $a = \frac{1}{2}$, with corresponding function value

$f(a) = 2$. We have used dashed horizontal lines to indicate an ε-band about $f(a) = 2$ on the y-axis. We have indicated with dashed vertical lines the interval on the x-axis that is mapped to this ε-band about 2. Next consider $a = \frac{1}{8}$ and the corresponding function value $f(a) = 8$. The ε band about 8 is the same width as the ε-band about 2, but the interval about $a = \frac{1}{8}$ that is mapped to the ε band is much smaller than the interval about $a = \frac{1}{2}$. We see that, as we choose a closer and closer to zero, the largest δ for which $|x - a| < \delta$ implies $|f(x) - f(a)| < \varepsilon$ becomes smaller and smaller.

With this conceptual understanding, we turn to the proof that f is not uniformly continuous on $(0, 1)$. We must show that there exists $\varepsilon > 0$ such that, for all $\delta > 0$, there exist $x, a \in (0, 1)$ with $|x - a| < \delta$ but $|f(x) - f(a)| \geq \varepsilon$. Take $\varepsilon = 1$ and let $\delta > 0$ be given. If $\delta \geq 1$, set $x = \frac{1}{2}$ and $a = \frac{1}{4}$ and observe that $|x - a| < 1 \leq \delta$ but

$$|f(x) - f(a)| = \left|\frac{1}{\frac{1}{2}} - \frac{1}{\frac{1}{4}}\right| = |2 - 4| = 2 \geq \varepsilon.$$

If $\delta < 1$, set $x = \delta$ and $a = \frac{\delta}{2}$ and observe that $x, a \in (0, 1)$ with $|x - a| = \frac{\delta}{2} < \delta$. Furthermore,

$$\begin{aligned} |f(x) - f(a)| &= \left|\frac{1}{\delta} - \frac{2}{\delta}\right| \\ &= \frac{1}{\delta} \\ &> 1 = \varepsilon. \end{aligned}$$

Thus f is not uniformly continuous on $(0, 1)$.

Contrast the previous example with the next, in which we consider a function given by the same formula but on a different set.

Example 12.8. We know that $f\colon [1, \infty) \to \mathbb{R}$ given by $f(x) = \frac{1}{x}$ is continuous at every point of $[1, \infty)$. We prove that f is uniformly continuous on $[1, \infty)$.

Let $\varepsilon > 0$ be given. We must show that there exists $\delta > 0$ such that, for all $x, a \in [1, \infty)$ with $|x - a| < \delta$, $|f(x) - f(a)| < \varepsilon$. Observe,

$$\begin{aligned} |f(x) - f(a)| &= \left|\frac{1}{x} - \frac{1}{a}\right| \\ &= \frac{1}{ax}|x - a| \\ &\leq |x - a| \end{aligned}$$

because $x, a \geq 1$ implies $\frac{1}{ax} \leq 1$. Thus we may take $\varepsilon = \delta$.

Exercise 12.13. Show that $f(x) = x^2$ is not uniformly continuous on $\mathbb{R}$.

Exercise 12.14. Show that $f(x) = x^3$ is uniformly continuous on $[-2, 2]$.

12.2.3 The limit of a function

Let $D \subseteq \mathbb{R}$, let $f\colon D \to \mathbb{R}$, and let a be an accumulation point of D. Whether or not f is defined at a, and regardless of its value at a if it is defined, we can ask whether f has a *limit* as x approaches a.

Definition 12.4. Let $D \subseteq \mathbb{R}$, let $f\colon D \to \mathbb{R}$, and let a be an accumulation point of D. We say that f has **limit** L at a if, for all sequences $\{x_n\}$ in $D \setminus \{a\}$ converging to a, $\{f(x_n)\}$ converges to L. When this condition is satisfied, we write $\lim_{x\to a} f(x) = L$.

This definition closely resembles the definition of continuity at a with two differences:

1. a can never appear as a term in the sequence $\{x_n\}$, and
2. f needn't be defined at a, and, even if it is, L need not equal $f(a)$.

Of course, if f is continuous at a, it has a limit at a, and this limit equals $f(a)$.

Example 12.9. To illustrate the definition, we consider an example of a function that is not defined at 2 but whose limit at 2 exists. Consider $f\colon \mathbb{R}\setminus\{2\} \to \mathbb{R}$ defined by

$$f(x) = \frac{x^3 - 4x}{x - 2}.$$

Our graphical and numerical explorations at the beginning of this chapter suggest that $\lim_{x\to 2} f(x)$ exists and equals 8. To prove this claim, consider an arbitrary sequence $\{x_n\}$ with limit 2 but with no term equal to 2. Thus for all n, $x_n - 2 \neq 0$. Consider $\{f(x_n)\}$.

$$\begin{aligned} f(x_n) &= \frac{x_n^3 - 4x_n}{x_n - 2} \\ &= \frac{x_n(x_n + 2)(x_n - 2)}{x_n - 2} \\ &= x_n(x_n + 2) \qquad \text{(because } x_n - 2 \neq 0\text{).} \end{aligned}$$

Because $\lim_{n\to\infty} x_n = 2$, by the limit laws, $\lim_{n\to\infty}(x_n + 2) = 4$ and

$$\lim_{n\to\infty} x_n(x_n + 2) = 2 \cdot 4 = 8.$$

Thus $\lim_{n\to\infty} f(x_n) = 8$. Therefore f has limit 8 at $a = 2$.

Example 12.10. Our second example illustrates how we prove that a function does not have a limit at a certain point. Consider $f\colon \mathbb{R} \setminus \{0\} \to \mathbb{R}$ defined by

$$f(x) = \begin{cases} 1 & x > 0 \\ -1 & x < 0. \end{cases}$$

We show that $\lim_{x\to 0} f(x)$ does not exist. Negating the condition in Definition 12.4, we see that we must find a sequence $\{x_n\}$ in $\mathbb{R}\setminus\{0\}$ such that $\lim_{n\to\infty} x_n =$

0 but $\lim_{n\to\infty} f(x_n)$ does not exist. We must be careful about how we select our sequence. For example, if we try $x_n = \frac{1}{n}$ for $n \in \mathbb{N}$, $f(x_n) = 1$ for every n, and in this case the sequence $\{f(x_n)\}$ *does* have a limit.

Let's think more carefully about the behavior of this function. We believe its limit at 0 does not exist because the function jumps from a value of -1 to the left of 0 to a value of 1 to the right of 0. Because the problem with this function is that its values on either side of 0 are very different, we should take a sequence that has infinitely many terms on either side of 0. Thus we consider $x_n = \frac{(-1)^n}{n}$ for $n \in \mathbb{N}$. Certainly $\lim_{n\to\infty} x_n = 0$. Furthermore, $f(x_n) = (-1)^n$. Because $\{f(x_n)\}$ is divergent, $\lim_{x\to 0} f(x)$ does not exist.

Exercise 12.15. Consider $f\colon \mathbb{R} \to \mathbb{R}$ defined by

$$f(x) = \begin{cases} x & x < -1 \\ x^2 & x \geq -1. \end{cases}$$

(a) Let $x_n = -\frac{n}{n+1}$. Find $\lim_{n\to\infty} x_n$ and $\lim_{n\to\infty} f(x_n)$. Does this calculation allow you to determine whether $\lim_{x\to -1} f(x)$ exists?

(b) Determine, with proof, whether $\lim_{x\to -1} f(x)$ exists.

Our definition of the limit of a function and the limit laws for sequences (Proposition 8.3) easily give us the limit laws for functions.

Proposition 12.2. *Let $D \subseteq \mathbb{R}$, let $f, g\colon D \to \mathbb{R}$, and let a be an accumulation point of D. Suppose $\lim_{x\to a} f(x) = L$ and $\lim_{x\to a} g(x) = K$. Then*

(a) for any $A \in \mathbb{R}$, $\lim_{x\to a}(Af)(x)$ exists and equals AL.

(b) $\lim_{x\to a}(f+g)(x)$ exists and equals $L + K$.

(c) $\lim_{x\to a}(fg)(x)$ exists and equals LK.

(d) if $K \neq 0$, $\lim_{x\to a} \left(\frac{f}{g}\right)(x)$ exists and equals $\frac{L}{K}$.

Proof. We give the proof of (c) and leave the rest for the reader. Let $\{x_n\}$ be any sequence in $D \setminus \{a\}$ with $\lim_{n\to\infty} x_n = a$. We must show that $\lim_{n\to\infty}(fg)(x_n)$ exists and equals LK. Now, $(fg)(x_n) = f(x_n)g(x_n)$. Because $\lim_{n\to\infty} f(x_n) = L$ and $\lim_{n\to\infty} g(x_n) = K$, by the limit laws for sequences, $\lim_{n\to\infty} f(x_n)g(x_n)$ exists and equals LK. The result follows. The proofs of the other parts of the proposition are just as easy and are left as exercises. □

Clearly $\lim_{x\to a} x = a$ and $\lim_{x\to a} c = c$. With these basic facts and Proposition 12.2, we can deal easily with many limits. We will use this proposition heavily in the next chapter when we want to take limits of difference quotients, which are often rational expressions but which are not defined at the point at which we wish to take the limit.

In Theorem 12.1, we saw that sequential continuity is equivalent to a certain ε-δ condition. Not surprisingly, the condition in Definition 12.4 is also equivalent to a certain ε-δ condition. We leave the proof to the reader.

Proposition 12.3. *Let $f\colon D \subseteq \mathbb{R} \to \mathbb{R}$ and let a be an accumulation point of D. f has limit L at a if and only if, for every $\varepsilon > 0$ there exists $\delta > 0$ such that, for all $x \in D$ with $0 < |x-a| < \delta$, $|f(x) - L| < \varepsilon$.*

Exercise 12.16. Prove Proposition 12.3.

12.3 Important Theorems

12.3.1 The Intermediate Value Theorem

We begin with a theorem about continuous functions that will probably strike you as obvious from your extensive experience with continuous functions from calculus. Although this result is not surprising, its proof is important because it is *constructive*; it does not simply prove that an object exists, it tells us how to find it.

Theorem 12.2 (Intermediate Value Theorem (IVT)). *Suppose $a < b$ and let f be a continuous real-valued function on $[a,b]$. If y is any real number between $f(a)$ and $f(b)$, then there exists $c \in [a,b]$ such that $f(c) = y$.*

Proof. If $f(a) = f(b)$, the only possibility for y is this common value for $f(a)$ and $f(b)$, and there is nothing to prove, for we may take $c = a$ or $c = b$. Thus we suppose $f(a) \neq f(b)$. We suppose $f(a) < f(b)$; the case in which $f(a) > f(b)$ is handled nearly identically and is left for the reader.

Take y such that $f(a) < y < f(b)$. We use a divide-and-conquer approach to inductively construct two sequences $\{a_n\}$ and $\{b_n\}$. The former will consist of points at which the value of the function is less than or equal to y and the latter will consist of points at which the function is greater than y. Let $a_0 = a$, $b_0 = b$. Then $f(a_0) \leq y$, $f(b_0) > y$, and $|a_0 - b_0| = |a - b|$. Proceed inductively; having selected a_k and b_k for $0 \leq k \leq n$, consider $[a_n, b_n]$ and its midpoint $m = \frac{a_n + b_n}{2}$. If $f(m) \leq y$, set $a_{n+1} = m$ and $b_{n+1} = b_n$. If $f(m) > y$, set $a_{n+1} = a_n$ and $b_{n+1} = m$. In either case, $f(a_{n+1}) \leq y$, $f(b_{n+1}) > y$, and $|a_{n+1} - b_{n+1}| = \frac{1}{2}|a_n - b_n| = \frac{1}{2^{n+1}}|a - b|$.

Now, because $\{a_n\}$ is a non-decreasing sequence bounded above by b, it converges to some real number $c \in [a,b]$. Also, because $\{b_n\}$ is a non-increasing sequence bounded below by a, it converges to some real number $c' \in [a,b]$. We claim $c = c'$. Indeed,

$$\begin{aligned} |c - c'| &= \lim_{n\to\infty} |a_n - b_n| \\ &\leq \lim_{n\to\infty} \frac{1}{2^n}|a - b| \\ &= 0, \end{aligned}$$

and $c' = c$.

Now consider $f(c)$. Because f is continuous,

$$f(c) = \lim_{n\to\infty} f(a_n) = \lim_{n\to\infty} f(b_n).$$

We claim $f(c) = y$. Because $f(a_n) \leq y$ and non-strict inequalities are preserved when taking limits, $f(c) \leq y$. Because $f(b_n) > y$ and non-strict inequalities are preserved when taking limits, $f(c) \geq y$. The conclusion follows. □

This theorem is not only of theoretical significance; it has an important practical application to finding roots. Recall that, if p is a polynomial, c is a *root* of p if $p(c) = 0$. If p has a simple form, we may be able to use techniques from algebra such as factoring to find its roots. Most of the time, however, it is difficult or impossible to find the roots algebraically. The Intermediate Value Theorem can be used in this situation to show that a function has a root on a certain interval.

Example 12.11. Consider $p(x) = x^5 + x^3 + 2x - 2$. We claim that p has a root in the interval $[0, 1]$. Observe that $p(0) = -2$ and $p(1) = 2$. Because p is continuous on all of $\mathbb{R}$, it is continuous on $[0, 1]$. By the Intermediate Value Theorem, because $p(0) < 0 < p(1)$, there exists $c \in (0, 1)$ such that $p(c) = 0$.

Exercise 12.17. Show that the equation $x^7 - 4x^3 + x = 2$ has at least one solution on the interval $[0, 2]$.

12.3.2 Developing a root-finding algorithm from the proof of the IVT

The Intermediate Value Theorem can do much more than just show that a function has a root on a certain interval $[a, b]$. The proof we have given actually gives us an algorithm we can implement to find the root. This algorithm, like several we have previously seen, is a divide-and-conquer algorithm that constructs two sequences $\{a_n\}$ and $\{b_n\}$ whose limit is the desired point. If we execute the algorithm through step N, either of the numbers a_N or b_N can be used as our approximation of the desired point. Furthermore, because that point c is between a_n and b_n for all n and $|a_n - b_n| = \frac{1}{2^n}|a - b|$, we get an upper bound on the error of our approximation.

Let's implement the algorithm of the proof of the Intermediate Value Theorem in Python to find the root of a specific polynomial.

Example 12.12. Consider $p(x) = 2x^5 + x^4 - x^3 + 2x^2 - x - 2$. Because $p(0) = -2$, $p(1) = 1$, and p is continuous on $[0, 1]$, by the Intermediate Value Theorem, there exists $c \in [0, 1]$ such that $p(c) = 0$. We will approximate this c.

```
# Construct sequences a and b. a will contain points where p is
    negative and b will contain points where it is positive.
a = [0.0] #left endpoint of initial interval
b = [1.0] #right endpoint of initial interval

# Loop at most 1000 times
for n in range(1,1000):
    x = (a[n-1]+b[n-1])/2 #midpoint of current interval
```

```
    if 2*x**5+x**4-x**3+2*x**2-x-2 <= 0: #p is still not
        positive at the midpoint
        # add x to a. Leave term of b unchanged
        a.append(x)
        b.append(b[n-1])
    else: #p is positive at the midpoint
        # add x to b. Leave term of a unchanged
        a.append(a[n-1])
        b.append(x)

    # check that something meaningful has changed. If it
    # hasn't, we've exceeded the precision of our floating point
        variables
    if a[n] == a[n-1] and b[n] == b[n-1]:
        print("Floating point precision exceeded.")
        break

    # probably won't happen, but it would be a stopping point if
        it did
    if a[n] == b[n]:
        print("a[n] = b[n]; this common value is our limit")
        break

# Approximation complete. Output results
print("Completed in "+str(n)+" iterations.")
print("Max a[n] = "+ str(a[n]))
print("Min b[n] = "+ str(b[1n]))
```

Here's the output:

```
Floating point precision exceeded.
Completed in 54 iterations.
Max a[n] = 0.917234048884
Min b[n] = 0.917234048884
```

We use 0.917234048884 as our estimate for the root of the polynomial.

You should compare this algorithm to the one we used in Example 9.4. The algorithm used to approximate $\sqrt{2}$ is exactly what we would get if we applied the algorithm of the proof of the IVT to find the root of the continuous function $f(x) = x^2 - 2$ on the interval $[1, 2]$.

Exercise 12.18. Between what two integers is the solution to $x^4 = 100$? Modify the code in Example 12.12 to approximate this solution.

12.3.3 Continuous functions on compact intervals

We embarked on our study of topological properties of $\mathbb{R}$ with the goal of proving that a continuous function on $[a, b]$ attains a maximum and minimum value.

In this subsection we prove this theorem and other theorems showing that continuous functions on compact sets have some very nice properties.

Proposition 12.4. *Let $K \subseteq \mathbb{R}$ be a non-empty compact set and suppose $f: K \to \mathbb{R}$ is continuous on K. Then f is bounded.*

Proof. We prove the contrapositive. Suppose f is not bounded. Then for every $n \in \mathbb{N}$, there exists $x_n \in K$ such that $|f(x_n)| > n$. Now, because K is compact, there exists a subsequence $\{x_{n_k}\}$ of $\{x_n\}$ and a point $a \in K$ such that $\{x_{n_k}\}$ converges to a. But because the sequence $\{f(x_n)\}$ is unbounded, $\{f(x_{n_k})\}$ is unbounded as well and hence does not converge to $f(a)$. Thus f is not continuous at a. □

Theorem 12.3 (Extreme Value Theorem)**.** *Let $K \subseteq \mathbb{R}$ be a non-empty compact set and suppose $f: K \to \mathbb{R}$ is continuous on K. Then there exist $c, d \in K$ such that $f(d) \leq f(x) \leq f(c)$ for all $x \in K$. In other words, f attains a maximum and a minimum value on K.*

Proof. We prove that f attains a maximum value and leave the proof that f attains a minimum value as an exercise.

Let $f(K) = \{y \in \mathbb{R} : y = f(x) \text{ for some } x \in K\}$. Because K is not empty, $f(K)$ is not empty. By Proposition 12.4, f is a bounded function, and so $f(K)$ is a bounded subset of $\mathbb{R}$. Thus there exists a real number M such that M is the supremum of $f(K)$. We show that there exists $c \in K$ with $f(c) = M$. For each $n \in \mathbb{N}$, $M - \frac{1}{n}$ is not an upper bound for $f(K)$ and so there exists $y_n = f(x_n) \in f(K)$ such that $M - \frac{1}{n} < f(x_n) \leq M$. Observe that, by construction, $\{y_n\} = \{f(x_n)\}$ is convergent, with limit M. Consider $\{x_n\}$. Because $x_n \in K$ and K is compact, there exists a subsequence $\{x_{n_k}\}$ that converges to a point c of K. Consider $\{f(x_{n_k})\}$. Because f is continuous, $\{f(x_{n_k})\}$ converges to $f(c)$. On the other hand, because $\{f(x_n)\}$ converges to M and $\{f(x_{n_k})\}$ is a subsequence, $\{f(x_{n_k})\}$ converges to M as well. Thus $f(c) = M$. □

You can easily use the ideas from the previous two proofs to prove the following theorem.

Theorem 12.4. *Let $K \subseteq \mathbb{R}$ be a non-empty compact set and let $f: K \to \mathbb{R}$. If f is continuous on K, then $f(K)$ is compact.*

Exercise 12.19. Prove Theorem 12.4.

12.4 Problems

1. For each of the following statements, either give a proof if it is true or give a counterexample if it is false.

   (a) Let $f, g: \mathbb{R} \to \mathbb{R}$ and suppose $f + g$ is continuous at 0. Then f and g are both continuous at 0.

(b) Let $f, g\colon \mathbb{R} \to \mathbb{R}$ and suppose fg is continuous at 0. Then f and g are both continuous at 0.

(c) Let $f\colon \mathbb{R} \to \mathbb{R}$ and suppose cf is continuous at 0 for some fixed constant c. Then f is continuous at 0.

(d) Let $f\colon \mathbb{R} \to \mathbb{R}$ and suppose cf is continuous at 0 for all $c \in \mathbb{R}$. Then f is continuous at 0.

2. In this chapter, when we have considered the continuity of functions $f\colon D \to \mathbb{R}$, usually the set D has been an open or closed interval. But the definition of continuity places no restrictions on D. Suppose $D = \{a_1, \ldots, a_N\}$, i.e., D is a finite subset of $\mathbb{R}$. Which functions $f\colon D \to \mathbb{R}$ are continuous on D?

3. Each function below is known to be continuous on the given domain. Determine (with proof) whether each function is uniformly continuous on its domain.

   (a) $f\colon (0,1) \to \mathbb{R}$ given by $f(x) = x^2$.

   (b) $f\colon \mathbb{R} \to \mathbb{R}$ given by $f(x) = ax + b$ for non-zero constants a and b.

   (c) $f\colon (1,\infty) \to \mathbb{R}$ given by $f(x) = \sqrt{x}$.

   (d) $f\colon (0,1) \to \mathbb{R}$ given by $f(x) = \frac{1}{\sqrt{x}}$.

4. Consider the *Dirichlet function* $D\colon \mathbb{R} \to \mathbb{R}$ defined by

$$D(x) = \begin{cases} 1 & \text{if } x \text{ is rational} \\ 0 & \text{if } x \text{ is irrational.} \end{cases}$$

   Find all points of $\mathbb{R}$ at which D is continuous. (If you have not yet done Problem 2 in Chapter 11, you should do so first.)

5. Give an example of each of the following or explain why no such example exists.

   (a) A function $f\colon [0,1] \to \mathbb{R}$ that is unbounded.

   (b) A bounded function $f\colon [-1,1] \to \mathbb{R}$ that attains a maximum but not a minimum value.

   (c) A continuous function $f\colon (-1,1) \to \mathbb{R}$ that is unbounded.

   (d) A continuous function $f\colon (-1,2) \to \mathbb{R}$ that attains a minimum value but not a maximum value.

   (e) A discontinuous function $f\colon [-1,1] \to \mathbb{R}$ that attains both a maximum and a minimum value.

   (f) A function $f\colon [0,1] \to \mathbb{R}$ for which $f(0) = -1$, $f(1) = 1$, but for which there is no c such that $f(c) = 0$.

(g) A discontinuous function $f\colon [0,1] \to \mathbb{R}$ for which $f(0) = -1$, $f(1) = 1$, and for which, for all $y \in [-1,1]$ there exists $c \in [0,1]$ such that $f(c) = y$.

6. Consider the function $f\colon \mathbb{R} \to \mathbb{R}$ defined by

$$f(x) = \begin{cases} \sin\left(\frac{\pi}{x}\right) & x \neq 0 \\ 0 & x = 0. \end{cases}$$

In this problem, you may freely use the properties of the sine function you learned in pre-calculus and calculus.

(a) Consider the sequence $\{x_n\}$ given by $x_n = \frac{1}{n}$. Does $\lim_{n\to\infty} f(x_n)$ exist? Can you conclude that f is continuous at 0?

(b) Find a sequence $\{x_n\}$ with $\lim_{n\to\infty} x_n = 0$ and $\lim_{n\to\infty} f(x_n) = 1$.

(c) Fix $c \in [-1,1]$. Find a sequence $\{x_n\}$ with $\lim_{n\to\infty} x_n = 0$ and $\lim_{n\to\infty} f(x_n) = c$.

(d) Define $g\colon \mathbb{R} \to \mathbb{R}$ by

$$g(x) = \begin{cases} x\sin\left(\frac{\pi}{x}\right) & x \neq 0 \\ 0 & x = 0. \end{cases}$$

Is g continuous at 0?

Programming Project. Let p be a polynomial of odd degree.

1. Prove that p has at least one real root.

2. Write code for a function that takes as input an odd-degree polynomial p and returns as output an interval on which p has a root. Although such a p is guaranteed to have a root, it could be very large. Thus you may want to build into your function a way to terminate the process if by some reasonable point no such interval is found.

Chapter 13

Differentiation

An introduction to analysis would seem incomplete if we didn't discuss differentiation. It is not, however, our aim to cover every familiar theorem from calculus. Rather, we think of the current chapter as an illustration of the way in which the theory of limits allows us to put calculus on a solid theoretical foundation.

13.1 Definition and First Examples

We begin with the familiar definition of differentiability from first-semester calculus.

Definition 13.1. Let I be an interval in $\mathbb{R}$. A function $f: I \to \mathbb{R}$ is **differentiable** at $a \in I$ if

$$\lim_{h \to 0} \frac{f(a+h) - f(a)}{h} \tag{13.1}$$

exists. When the limit exists, we denote it by $f'(a)$.

Definition 13.2. Suppose $f: I \to \mathbb{R}$ is differentiable at every point of I. Define $f': I \to \mathbb{R}$ by the rule $x \mapsto f'(x)$. We call f' the **derivative** of f.

Even though you have used Definition 13.1 many times before, we do several examples, paying special attention to the ways in which we use the theorems about limits established in earlier chapters.

Example 13.1. We show that $f: \mathbb{R} \to \mathbb{R}$ defined by $f(x) = x^2$ is differentiable at every $x \in \mathbb{R}$ with derivative $f'(x) = 2x$. Our definition says that we must show that, for every x, $\lim_{h \to 0} \frac{f(x+h)-f(x)}{h}$ exists. Recall that when we take the limit of a function at a point, we never consider the value of the function at that point. In this case, we are taking the limit of a function of h at 0. Thus we will never consider the value of the difference quotient for $h = 0$. For $h \neq 0$,

$$\begin{aligned} \frac{f(x+h)-f(x)}{h} &= \frac{(x+h)^2 - x^2}{h} \\ &= \frac{2xh + h^2}{h} \\ &= 2x + h. \end{aligned}$$

Using the limit laws for functions (Proposition 12.2), we obtain

$$\lim_{h \to 0} \frac{f(x+h)-f(x)}{h} = \lim_{h \to 0} (2x + h) = 2x.$$

Thus f is differentiable at x with $f'(x) = 2x$.

Example 13.2. (A non-differentiable function.) We consider the canonical example of a function that is continuous everywhere but fails to be differentiable at a point. Take $f\colon \mathbb{R} \to \mathbb{R}$ defined by $f(x) = |x|$. We claim that f is not differentiable at 0. Indeed, consider the difference quotient with $a = 0$. We easily find that

$$\frac{f(0+h)-f(0)}{h} = \frac{|h|}{h} = \begin{cases} 1 & h > 0 \\ -1 & h < 0 \end{cases}.$$

In Example 12.10, we showed that this function does not have a limit at 0, hence f is not differentiable at 0.

Exercise 13.1. Prove that $f\colon \mathbb{R} \to \mathbb{R}$ defined by $f(x) = |x|$ is continuous on $\mathbb{R}$.

Exercise 13.2. Let $f(x) = \frac{1}{\sqrt{x}}$. Show that f is differentiable at every $x \in (0, \infty)$.

Exercise 13.3. Let $f(x) = \sqrt{x}$. Show that f is not differentiable at 0.

If we replace $a + h$ with x in Definition 13.1, we immediately obtain a condition equivalent to differentiability at a point.

Proposition 13.1. *Let I be an interval in $\mathbb{R}$ and let $a \in I$. $f\colon I \to \mathbb{R}$ is differentiable at a if and only if $\lim_{x \to a} \frac{f(x)-f(a)}{x-a}$ exists.*

Although its proof is trivial, this proposition is worth stating because the different conditions require us to do different algebra as we simplify the difference quotient, and sometimes one kind of algebra seems easier than another. When we use the form of the difference quotient in Definition 13.1, we tend to expand expressions, whereas when we use the form of the difference quotient in Proposition 13.1 we tend to factor. We consider an example.

Example 13.3. We use the condition in Proposition 13.1 to show that $f\colon \mathbb{R} \to \mathbb{R}$ defined by $f(x) = x^4$ is differentiable at every $a \in \mathbb{R}$. Because we will take the limit of the difference quotient as x approaches a, we will never consider $x = a$, and so $x - a \neq 0$ in the difference quotient. We find

$$\begin{aligned}\frac{x^4 - a^4}{x - a} &= \frac{(x^2 - a^2)(x^2 + a^2)}{x - a} \\ &= \frac{(x - a)(x + a)(x^2 + a^2)}{x - a} \\ &= (x + a)(x^2 + a^2).\end{aligned}$$

Then by the limit laws for functions,

$$\lim_{x \to a} \frac{x^4 - a^4}{x - a} = \lim_{x \to a} (x + a)(x^2 + a^2) = (2a)(2a^2) = 4a^3,$$

and f is indeed differentiable at every point of $\mathbb{R}$.

We certainly do not want to have to take the limit of a difference quotient every time we want to find a derivative. We take the same approach we took in calculus. First, we use the limit definition to find derivatives of some basic functions. Second, we prove general rules for differentiating various combinations of these basic functions. The next proposition accomplishes the first step.

Proposition 13.2. *Let $c \in \mathbb{R}$ and let $n \in \mathbb{N}$.*

(a) Define $f\colon \mathbb{R} \to \mathbb{R}$ by the rule $f(x) = c$. Then f is differentiable on $\mathbb{R}$ with $f'(x) = 0$ for every x.

(b) (power rule) Define $f\colon \mathbb{R} \to \mathbb{R}$ by $f(x) = x^n$. Then f is differentiable on $\mathbb{R}$ with $f'(x) = nx^{n-1}$.

Proof. Part (a) is immediate because, for all x, the difference quotient is the constant function 0, hence has limit 0.

For part (b), we will use the Binomial Theorem to expand the numerator of the difference quotient.

$$\begin{aligned}\frac{(x + h)^n - x^n}{h} &= \frac{\sum_{k=0}^{n} \binom{n}{k} x^{n-k} h^k - x^n}{h} \\ &= \frac{x^n + nx^{n-1}h + \sum_{k=2}^{n} \binom{n}{k} x^{n-k} h^k - x^n}{h}. \qquad (13.2)\end{aligned}$$

If $n = 1$, $\sum_{k=2}^{n} \binom{n}{k} x^{n-k} h^k$ is empty and is thus equal to zero. We may now simplify the quotient, recalling that because we are taking a limit as h tends to 0, in the difference quotient, h is never allowed to be 0. Thus

$$\begin{aligned}(13.2) &= \frac{nx^{n-1}h + h^2 \sum_{k=2}^{n} \binom{n}{k} x^{n-k} h^{k-2}}{h} \\ &= nx^{n-1} + h \sum_{k=2}^{n} \binom{n}{k} x^{n-k} h^{k-2}.\end{aligned}$$

We find that

$$\lim_{h\to 0}\frac{(x+h)^n - x^n}{h} = \lim_{h\to 0}\left(nx^{n-1} + h\sum_{k=2}^{n}\binom{n}{k}x^{n-k}h^{k-2}\right) = nx^{n-1},$$

as claimed. □

13.2 Properties of Differentiable Functions and Rules for Differentiation

Before we can prove the familiar rules for differentiating sums, products, compositions, etc., we need to know some general facts about differentiable functions. Our first proposition gives a useful equivalent condition for differentiability.

Proposition 13.3. *$f\colon I \to \mathbb{R}$ is differentiable at $a \in I$ if and only if there exists a real number L and a function ϕ defined in some deleted neighborhood $(-\delta, \delta) \setminus \{0\}$ of zero such that, for all $h \in (-\delta, \delta) \setminus \{0\}$,*

$$f(a+h) = f(a) + Lh + \phi(h) \tag{13.3}$$

where

$$\lim_{h\to 0}\frac{\phi(h)}{h} = 0. \tag{13.4}$$

We will call a function ϕ with the property (13.4) an *error function.*

Proof. Suppose first that f is differentiable at a. We may write

$$f(a+h) = f(a) + f'(a)h + \left[\frac{f(a+h)-f(a)}{h} - f'(a)\right]h.$$

Consider

$$\phi(h) = \left[\frac{f(a+h)-f(a)}{h} - f'(a)\right]h.$$

Because $\lim_{h\to 0}\frac{f(a+h)-f(a)}{h} = f'(a)$, $\lim_{h\to 0}\frac{\phi(h)}{h} = 0$. Thus $f(a+h)$ can be written in the desired form, and the real number L is $f'(a)$.

On the other hand, if $f(a+h)$ can be written in the form (13.3) for some real L, then

$$\frac{f(a+h)-f(a)}{h} - L = \frac{\phi(h)}{h}.$$

Because the right-hand side tends to 0 as h tends to 0, $\lim_{h\to 0}\frac{f(a+h)-f(a)}{h} = L$. Therefore, f is differentiable at a, with $f'(a) = L$. □

Proposition 13.3 can be rephrased using geometric language. If we change notation slightly, letting $x = a + h$, then the proposition says that f is differentiable at a if and only if

$$f(x) = f(a) + f'(a)(x-a) + \phi(x-a), \tag{13.5}$$

where $\lim_{x\to a} \frac{\phi(x-a)}{x-a} = 0$. The equation $y = f(a) + f'(a)(x-a)$ is the equation of the line tangent to the graph of $y = f(x)$ at the point $(a, f(a))$. Thus Proposition 13.3 says that a function is differentiable if and only if there is a linear function that approximates f so well near a that the error of the approximation is small even when compared to the distance from x to a. This interpretation of differentiability coincides well with the geometric intuition we develop for differentiable functions in first-year calculus.

Before we use Proposition 13.3 to prove other theorems about differentiable functions, let's see how it could be used to prove that a specific function is differentiable.

Example 13.4. Although we already know the result, we use the condition in Proposition 13.3 to show that $f\colon \mathbb{R} \to \mathbb{R}$ defined by $f(x) = x^3$ is differentiable at every $a \in \mathbb{R}$. Observe,

$$\begin{aligned} f(a+h) &= (a+h)^3 \\ &= a^3 + 3a^2h + 3ah^2 + h^3. \end{aligned}$$

This last expression is of the form $f(a) + Lh + \phi(h)$ for $\phi(h) = 3ah^2 + h^3$. We will have proved the differentiability of f if we show that ϕ is an error function. Because

$$\lim_{h\to 0} \frac{\phi(h)}{h} = \lim_{h\to 0} \frac{3ah^2 + h^3}{h} = \lim_{h\to 0} (3ah + h^2) = 0,$$

the result follows.

Exercise 13.4. Use Proposition 13.3 to prove that $f\colon \mathbb{R} \to \mathbb{R}$ defined by $f(x) = 2x^2 + 3x + 1$ is differentiable at every $a \in \mathbb{R}$.

When you visualize a differentiable function, you probably see a function with no breaks, holes, vertical asymptotes, corners, or cusps. In particular, you always visualize a continuous function. This intuition is correct, as we now prove.

Proposition 13.4. *If $f\colon I \to \mathbb{R}$ is differentiable at $a \in I$, then f is continuous at a.*

Proof. Let $\{x_n\}$ be a sequence in I with limit a. By Proposition 13.3 (using the form (13.5) of the equation),

$$f(x_n) = f(a) + f'(a)(x_n - a) + \phi(x_n - a), \tag{13.6}$$

where ϕ is an error function. Because $\lim_{n\to\infty} x_n = a$, $\lim_{n\to\infty}(x_n - a) = 0$. Therefore

$$\lim_{n\to\infty} \phi(x_n - a) = \lim_{n\to\infty} \frac{\phi(x_n - a)}{x_n - a}(x_n - a) = 0.$$

Thus (13.6) has limit $f(a)$, and so f is continuous at a. □

We are finally ready to prove the familiar rules for differentiation.

Proposition 13.5. *Let $f, g\colon I \to \mathbb{R}$ be differentiable at some $x \in I$, and let c be a real constant.*

(a) *cf is differentiable at x, with $(cf)'(x) = cf'(x)$.*

(b) *$f + g$ is differentiable at x, with $(f+g)'(x) = f'(x) + g'(x)$.*

(c) *(product rule) fg is differentiable at x, with*

$$(fg)'(x) = f'(x)g(x) + f(x)g'(x).$$

(d) *(quotient rule) If $g(x) \neq 0$, $\frac{f}{g}$ is differentiable at x, with*

$$\left(\frac{f}{g}\right)'(x) = \frac{f'(x)g(x) - f(x)g'(x)}{g(x)^2}.$$

Proof. We prove (c) and leave the proofs of the remaining parts as an exercise. We must prove that $\lim_{h\to 0} \frac{(fg)(x+h)-(fg)(x)}{h}$ exists. Observe,

$$\begin{aligned}
\frac{(fg)(x+h) - (fg)(x)}{h} &= \frac{f(x+h)g(x+h) - f(x)g(x)}{h} \\
&= \frac{f(x+h)g(x+h) - f(x)g(x+h)}{h} \\
&\quad + \frac{f(x)g(x+h) - f(x)g(x)}{h} \\
&= \frac{f(x+h) - f(x)}{h} g(x+h) + f(x)\frac{g(x+h) - g(x)}{h}.
\end{aligned}$$

Because g is differentiable at x, g is continuous at x, and so $\lim_{h\to 0} g(x+h) = g(x)$. We may use the limit laws for functions to conclude that the last sum above has limit

$$f'(x)g(x) + f(x)g'(x).$$

Thus the product rule is established. □

Exercise 13.5. Prove parts (a), (b), and (d) of Proposition 13.5.

As an application of these differentiation rules, let's prove the power rule for negative integers. *We do not yet know this proposition.* So far, we have only proved the power rule for natural numbers. That proof used the Binomial Theorem and can not be extended to negative integers.

Proposition 13.6. *Let $n \in \mathbb{Z}$ with $n < 0$. Consider $f\colon \mathbb{R} \setminus \{0\} \to \mathbb{R}$ defined by $f(x) = x^n$. Then f is differentiable at every point of $\mathbb{R}\setminus\{0\}$ with $f'(x) = nx^{n-1}$.*

Proof. Write $f(x) = \frac{1}{x^{-n}}$ and observe that $-n \in \mathbb{N}$. Thus f is a quotient of the everywhere differentiable constant function $p(x) = 1$ and the everywhere differentiable function $q(x) = x^{-n}$. Furthermore, by Proposition 13.2, $p'(x) = 0$

and $q'(x) = -nx^{-n-1}$. Because $q'(x) \neq 0$ if $x \neq 0$, by the quotient rule, $f = \frac{p}{q}$ is differentiable at every point of $\mathbb{R} \setminus \{0\}$ with derivative

$$\begin{aligned} f'(x) &= \frac{p'(x)q(x) - p(x)q'(x)}{[q(x)]^2} \\ &= \frac{0 - (-nx^{-n-1})}{[x^{-n}]^2} \\ &= \frac{nx^{-n-1}}{x^{-2n}} \\ &= nx^{2n-n-1} \\ &= nx^{n-1}, \end{aligned}$$

as claimed. □

We now know how to differentiate any algebraic combination of constants and power functions. In particular, we can differentiate any polynomial and any rational function at any point where its denominator is not zero. It is also important to know how to differentiate compositions of functions. The *chain rule* gives us a useful formula.

Proposition 13.7 (chain rule). *Let f and g be real-valued functions defined on subsets of $\mathbb{R}$. Suppose g is differentiable at a and f is differentiable at $b = g(a)$. Then $f \circ g$ is differentiable at a with $(f \circ g)'(a) = f'(g(a))g'(a)$.*

We can gain an intuition for the chain rule by thinking about linear functions. Suppose that on the side of a mountain, the temperature changes by $-0.1°F$ for every increase in elevation of 100 feet. Suppose that you are hiking, and that your elevation is increasing at a steady rate of 1500 feet per hour. Then you experience a decrease in temperature as you hike, with the temperature changing at a rate of

$$\frac{-0.1°F}{100\,\text{ft}} \cdot \frac{1500\,\text{ft}}{1\,\text{h}} = -1.5°F/\text{h}.$$

If f is the function that gives temperature as a function of elevation, and if g is the function that gives elevation as a function of time, then $f \circ g$ is the function that gives temperature as a function of time. The derivative of this function represents the rate of change of temperature with respect to time. We see that it is obtained by multiplying the derivative of f by the derivative of g, just as the chain rule suggests.

Exercise 13.6. Without using the chain rule, prove that if f and g are arbitrary linear functions, then, for any a, $(f \circ g)'(a) = f'(g(a))g'(a)$.

Now that the chain rule seems plausible, we give its proof.

Proof. Because g is differentiable at a and f is differentiable at b, there exist error functions ϕ_1 and ϕ_2 such that

$$\begin{aligned} g(a+h) &= g(a) + g'(a)h + \phi_1(h) \\ f(b+k) &= f(b) + f'(b)k + \phi_2(k). \end{aligned}$$

Taking $b = g(a)$ and $k = g'(a)h + \phi_1(h)$, we find

$$\begin{aligned}
&(f \circ g)(a+h) \\
&\quad = f(g(a+h)) \\
&\quad = f(g(a) + g'(a)h + \phi_1(h)) \\
&\quad = f(g(a)) + f'(g(a))(g'(a)h + \phi_1(h)) + \phi_2(g'(a)h + \phi_1(h)) \\
&\quad = f(g(a)) + f'(g(a))g'(a)h + [f'(g(a))\phi_1(h) + \phi_2(g'(a)h + \phi_1(h))].
\end{aligned}$$

Let $\phi(h)$ be the expression in square brackets. Our proof will be complete if we can show that ϕ is an error function. Observe,

$$\frac{\phi(h)}{h} = f'(g(a))\frac{\phi_1(h)}{h} + \frac{\phi_2(g'(a)h + \phi_1(h))}{h}. \tag{13.7}$$

We must consider the limit as h tends to 0. The first term on the right has limit 0 because ϕ_1 is an error function and $f'(g(a))$ is a constant. Consider the second term. If $g'(a)h + \phi_1(h)$ happens to equal 0 for some non-zero h, we will agree that $\phi_2(g'(a)h + \phi_1(h)) = 0$ because ϕ_2 has limit 0 at 0 and can sensibly be defined to have value 0 at 0. If $g'(a)h + \phi_1(h) \neq 0$, we write

$$\frac{\phi_2(g'(a)h + \phi_1(h))}{h} = \frac{\phi_2(g'(a)h + \phi_1(h))}{g'(a)h + \phi_1(h)} \cdot \frac{g'(a)h + \phi_1(h)}{h}.$$

Because $\lim_{h \to 0}(g'(a)h + \phi_1(h)) = 0$ and ϕ_2 is an error function, the first factor tends to 0 as h tends to 0. Because ϕ_1 is an error function, the second factor tends to the constant $g'(a)$. Thus the product tends to 0. We may now conclude that the second term in (13.7) also has limit 0. Therefore ϕ is an error function, as desired. □

Exercise 13.7. Let $f(x) = (x^2 + 1)^3$.

(a) Find $f'(x)$ without using the chain rule, by first expanding the binomial.

(b) Find $f'(x)$ using the chain rule.

(c) Reconcile your answers.

13.3 Applications of the Derivative

Perhaps the most important application of the derivative is to the problem of finding local extrema of functions.

Definition 13.3. Let $I \subseteq \mathbb{R}$ be an interval and let $c \in I$. We say that $f\colon I \to \mathbb{R}$ has a **local maximum** at c if there exists $\delta > 0$ such that, for all $x \in (c-\delta, c+\delta)$, $f(x) \leq f(c)$.

Exercise 13.8. Give an analogous definition of a *local minimum* of f on an interval I.

Figure 13.1: A function with local maxima that are not the global maximum on $[0, 7]$.

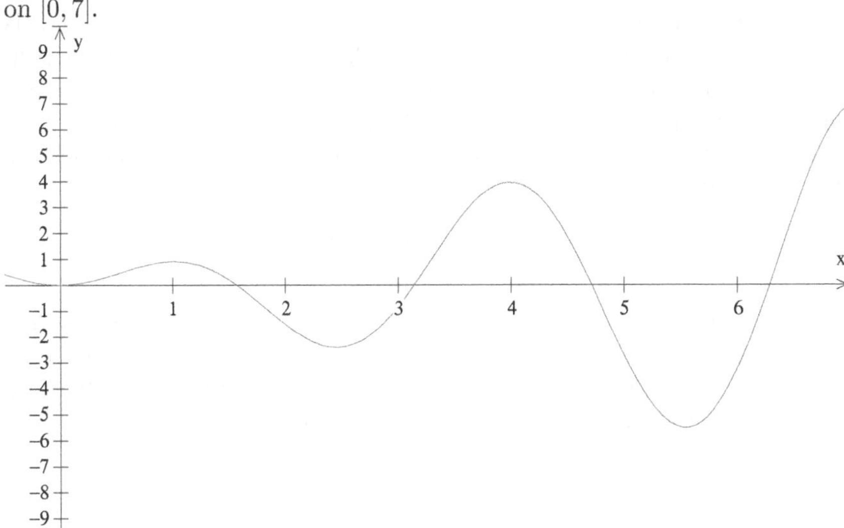

In words, f has a local maximum at c if there is some (possibly small) interval about c such that the value of the function at c equals or exceeds the value at every other point of the interval. A local maximum of a function defined on I need not be the (global) maximum of the function on I, and a function can have many local maxima. For example, the graph in Figure 13.1 shows a function on the interval $[0, 7]$ with two local maxima, neither of which is the maximum value of the function on $[0, 7]$. This example notwithstanding, finding local maxima is an important step in optimization problems because the maximum value of a continuous function on an interval $[a, b]$ occurs either at a local maximum or at one of the endpoints of the interval. So how do we find local maxima and minima? For differentiable functions, the answer is easy; we look for points at which the derivative is 0. The geometric intuition is that, at the tops of hills and the bottoms of valleys, the tangent line to a curve is horizontal, hence has slope 0. We prove this necessary condition for local extrema.

Proposition 13.8. *Let $I \subseteq \mathbb{R}$ be an open interval and let $f\colon I \to \mathbb{R}$ be differentiable on I. If f has a local maximum or a local minimum at $c \in I$, then $f'(c) = 0$.*

The converse is *not* true. For example, the function $f(x) = x^3$ is differentiable everywhere with derivative $f'(x) = 3x^2$. $f'(0) = 0$, but f does not have a local maximum or a local minimum at 0, for $f(x) < 0$ for $x < 0$ and $f(x) > 0$ for $x > 0$. Thus finding points at which a derivative is 0 merely gives us candidates for locations of local extrema.

Proof. We prove the proposition in the case in which c is the location of a local maximum and leave the other case as an exercise. By the definition of a local maximum, there exists $\delta > 0$ such that, for all $x \in (c-\delta, c+\delta)$, $f(x) \leq f(c)$. We consider difference quotients for f at c.

First suppose that $0 < h < \delta$. Then $f(c+h) - f(c) \leq 0$, and because h is positive, the difference quotient $\frac{f(c+h)-f(c)}{h}$ is less than or equal to 0 as well. Now, because $f'(c)$ exists, it can be obtained by taking the limit of this difference quotient for any sequence $\{h_n\}$ tending to 0. Take $\{h_n\}$ to be a sequence with limit 0 such that $0 < h_n < \delta$ for all n. Because non-strict inequalities are preserved when taking a limit,

$$f'(c) = \lim_{n\to\infty} \frac{f(c+h_n) - f(c)}{h_n} \leq 0.$$

Next, suppose that $-\delta < h < 0$. Then $f(c+h) - f(c) \leq 0$, and because h is negative, the difference quotient $\frac{f(c+h)-f(c)}{h}$ is greater than or equal to 0. Take $\{h_n\}$ to be a sequence with limit 0 such that $-\delta < h_n < 0$ for all n. Because non-strict inequalities are preserved when taking a limit,

$$f'(c) = \lim_{n\to\infty} \frac{f(c+h_n) - f(c)}{h_n} \geq 0.$$

Combining the two inequalities gives $f'(c) = 0$. □

Exercise 13.9. Prove Proposition 13.8 in the case in which c is the location of a local minimum of the function f.

We now easily obtain the key tool in optimization.

Theorem 13.1. *Let $f\colon [a,b] \to \mathbb{R}$ be a differentiable function. Then either f attains its maximum value at one of the endpoints of $[a,b]$, or it attains its maximum value at some $c \in (a,b)$ at which $f'(c) = 0$.*

Proof. We must first argue that f actually attains a maximum value on the interval $[a,b]$. Because f is differentiable on $[a,b]$, by Proposition 13.4, it is continuous on $[a,b]$. Because $[a,b]$ is a compact set, by the Extreme Value Theorem (Theorem 12.3), f does attain a maximum value on $[a,b]$.

If f does not attain its maximum at a or b, then there exists $c \in (a,b)$ such that $f(x) \leq f(c)$ for all $x \in (a,b)$. Because c is a point in the open interval (a,b), there certainly exists δ such that $(c-\delta, c+\delta) \subseteq (a,b)$, and so for all x in this smaller interval, $f(x) \leq f(c)$. Thus by definition, f has a local maximum at c, and by Proposition 13.8, $f'(c) = 0$. □

Of course, a similar result holds for the minimum of a function on a compact interval. We omit the proof, but we illustrate the use of these results. Problems like the following are standard in a calculus course.

Example 13.5. We find the maximum and minimum values of $f(x) = \frac{1}{3}x^3 - 4x + 1$ on $[-3, 5]$. First, we evaluate f at the endpoints. We find that $f(-3) = 4$ and $f(5) = \frac{68}{3}$. We now find the zeros of the derivative. Because

$$f'(x) = x^2 - 4 = (x-2)(x+2),$$

the only zeros of f' are at $x = \pm 2$. We now evaluate f itself at these points. We obtain $f(2) = -\frac{13}{3}$ and $f(-2) = \frac{19}{3}$. Comparing these four values, we see that the global maximum on the interval $[-3, 5]$ occurs at 5 and the global minimum occurs at 2.

13.4 Problems

The first few problems deal with *higher-order derivatives.* Let $I \subseteq \mathbb{R}$ be an interval and let $f\colon I \to \mathbb{R}$ be differentiable at all points of I. Consider $f'\colon I \to \mathbb{R}$. If f' is differentiable at some point a of I, we write $f''(a)$ or $f^{(2)}(a)$ for the value $(f')'(a)$. In a similar manner we can define derivatives of f of any positive order n, denoting the n-th derivative by $f^{(n)}$.

1. Let $n \in \mathbb{N}$ and consider the power function $f\colon \mathbb{R} \to \mathbb{R}$ defined by $f(x) = x^n$. Show that f has derivatives of all orders and find a formula for $f^{(k)}(x)$ for all $k \in \mathbb{N}$.

2. Find a function $f\colon \mathbb{R} \to \mathbb{R}$ that is differentiable at all points of $\mathbb{R}$ but whose second derivative fails to exist for at least one point of $\mathbb{R}$.

3. Consider $f\colon \mathbb{R} \setminus \{1\} \to \mathbb{R}$ defined by $f(x) = \frac{1}{1-x}$. Find $f^{(n)}(0)$ for all $n \in \mathbb{N}_0$. (We interpret $f^{(0)}$ to be the function f itself.)

4. Let $f(x) = \sum_{n=0}^{d} a_n x^n$ be a polynomial of degree d. Find and prove a formula relating $f^{(n)}(0)$ for $0 \leq n \leq d$ to the coefficients a_n.

5. Let $f, g\colon \mathbb{R} \to \mathbb{R}$ and suppose that f and g have derivatives of all orders at every point of $\mathbb{R}$. Let $n \in \mathbb{N}$. Find and prove a formula for $(fg)^{(n)}$, the n-th derivative of their product.

6. Fix a positive constant A. Find the dimensions of the rectangle with perimeter A having the largest possible area.

7. Let $n, m \in \mathbb{N}$ and let $f(x) = x^n(1-x)^m$. Find the maximum and minimum values of f on $[0, 1]$.

8. Give an example of each of the following or explain why no such function exists.

   (a) A differentiable function on $(-2, 2)$ that does not attain a maximum or a minimum value.

   (b) A differentiable function on $(-2, 2)$ that does attain a maximum and a minimum value.

   (c) A bounded, differentiable function on $(0, 1)$ whose derivative is unbounded.

   (d) A differentiable function on $[0, 1]$ whose derivative is unbounded.

9. Let $f\colon [a,b] \to \mathbb{R}$ be a differentiable function.
   (a) Suppose there exists $c \in (a,b)$ such that $f(c) > 0$. Prove that there exists $\delta > 0$ such that $f(x) > 0$ for every $x \in (c-\delta, c+\delta)$.
   (b) Prove Rolle's Theorem: If $f(a) = f(b) = 0$, there exists $c \in (a,b)$ such that $f'(c) = 0$.
   (c) Let $f\colon I \to \mathbb{R}$. Suppose that f is differentiable on I and that $f'(x) = 0$ for all $x \in I$. Show that f is constant.
10. Determine the exact number of roots of $f(x) = x^5 + 4x^3 + x - 10$ on $\mathbb{R}$ without finding them.

Programming Project. *Newton's method* gives a way to use the derivative to find a root of a function f. Here's the idea: Let x_0 be an initial guess of the location of the root. Find the equation of the line tangent to the graph of $y = f(x)$ at $(x_0, f(x_0))$. Take x_1 to be the x-intercept of this tangent line. x_1 is our next guess for the root. Proceed inductively; having obtained $x_0, \ldots, x_{n-1}$ for $n \geq 1$, let x_n be the x-intercept of the line tangent to the graph of $y = f(x)$ at $(x_{n-1}, f(x_{n-1}))$.

1. Let $f(x) = x^4 - 2$. If we take $x_0 = 2$, find x_1, x_2, and x_3 by hand.
2. Derive a recursive formula for the sequence $\{x_n\}$.
3. Write code to generate and plot the first 100 terms of the sequence $\{x_n\}$.
4. Does the sequence $\{x_n\}$ appear to be convergent? Can you prove your claim?
5. Newton's method does not always work to find a root. Give two examples in which Newton's method fails.

Chapter 14

Series, Part 2

In this chapter, we consider some of the subtleties of infinite series. Up to this point, most of our theorems about series have been about series with non-negative terms. For such series, because the partial sums form a non-decreasing sequence, convergence of the series is equivalent to the boundedness of the sequence of partial sums. Furthermore, whether or not the partial sums are bounded depends only on whether the terms of the series tend to zero quickly enough. In this chapter, we deal with arbitrary series of real (or complex) numbers. For such series, convergence depends not just on the absolute size of the terms but also on their order; if the positive and negative terms are arranged just so, even though they do not tend to zero very quickly, there may be additional cancellation so that the partial sums nonetheless have a limit.

14.1 Absolute and Conditional Convergence

In Chapter 10, any time we encountered a series $\sum a_n$ with both positive and negative terms, we also considered the associated non-negative series $\sum |a_n|$. We defined the notion of *absolute convergence*, which we now recall.

Definition 14.1. Let $\sum a_n$ be a series of real numbers. Then $\sum a_n$ **converges absolutely** if the series $\sum |a_n|$ converges.

We saw in Proposition 10.4 that if a series converges absolutely, then it converges. If you look back at all of our theorems in Chapter 10, you will see that any result that applies to series $\sum a_n$ with both positive and negative terms compares that series to the absolute series $\sum |a_n|$ and proves the convergence of the latter. (See, for example, the ratio test and the root test.) As we will see shortly, however, not all convergent series are absolutely convergent. We introduce some terminology for this situation.

Definition 14.2. Let $\sum a_n$ be a series of real numbers. Then $\sum a_n$ **converges conditionally** if the series converges but the associated absolute series $\sum |a_n|$ diverges.

Such conditionally convergent series are the subject of this chapter. Although they can be harder to work with than non-negative series, they have some fascinating properties.

14.1.1 The first example

Before we consider any general theorems about conditionally convergent series, we should give an example. Consider the series

$$\sum_{n=0}^{\infty} \frac{(-1)^n}{n+1}. \tag{14.1}$$

This series is not absolutely convergent; the associated absolute series $\sum_{n=0}^{\infty} \frac{1}{n+1}$ is the harmonic series, which we know to be divergent. We will show that $\sum_{n=0}^{\infty} \frac{(-1)^n}{n+1}$ is conditionally convergent.

By definition, in order to prove that this series converges, we must prove that the sequence $\{A_N\}$ of partial sums has a limit. In this case, $A_N = \sum_{n=0}^{N} \frac{(-1)^n}{n+1}$. Computing by hand rapidly becomes tedious, and it is hard to see from the first few partial sums whether they have a limit. Let's put our technology to work. We can easily generate and plot the first 50 partial sums. In this program, we use the recursive definition for the sequence $\{A_N\}$, which says that $A_0 = 1$ and $A_N = A_{N-1} + \frac{(-1)^N}{N+1}$ for $N \geq 1$.

```
# List A will contain the sequence of partial sums
A=[1.0] #The first partial sum is 1

for N in range(1,50):
    A.append(A[N-1]+(-1.0)**(N)/(N+1))

plot(A,".")
show()
```

Figure 14.1 shows the output. From this plot, we conjecture that the partial sums with even index form a sequence $\{A_{2N}\}$ that is non-increasing and convergent, and that the partial sums with odd index form a sequence $\{A_{2N+1}\}$ that is non-decreasing and convergent. Furthermore, these two subsequences of $\{A_N\}$ appear to have the same limit.

Now that we have these conjectures, we can explore the partial sums more closely with an eye toward proving the conjectures. First consider the partial sums A_{2N}. It will be especially convenient to relate each element of this subsequence to its predecessor. We find

$$\begin{aligned}
A_0 &= 1 \\
A_2 &= 1 + \left(-\frac{1}{2} + \frac{1}{3}\right) = A_0 - \frac{1}{6} \\
A_4 &= A_2 + \left(-\frac{1}{4} + \frac{1}{5}\right) = A_2 - \frac{1}{20},
\end{aligned}$$

Figure 14.1: First 50 partial sums of the alternating harmonic series.

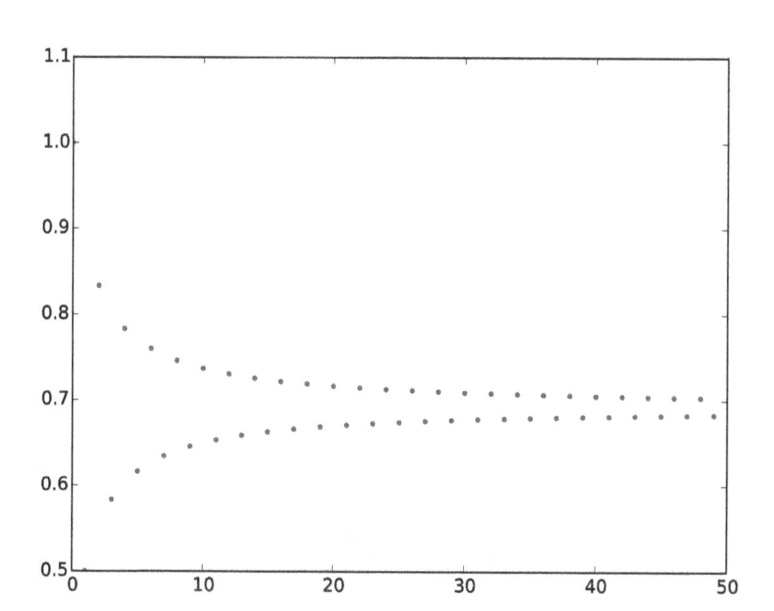

and, in general,

$$A_{2N} = A_{2N-2} - \frac{1}{2N(2N+1)}.$$

Thus the sequence $\{A_{2N}\}$ is decreasing. Next consider the partial sums $\{A_{2N+1}\}$. We find

$$\begin{aligned} A_1 &= 1 - \frac{1}{2} = \frac{1}{2} \\ A_3 &= \left(1 - \frac{1}{2}\right) + \left(\frac{1}{3} - \frac{1}{4}\right) = A_1 + \frac{1}{12} \\ A_5 &= A_3 + \left(\frac{1}{5} - \frac{1}{6}\right) = A_3 + \frac{1}{30}, \end{aligned}$$

and, in general,

$$A_{2N+1} = A_{2N-1} + \frac{1}{(2N+1)(2N+2)}.$$

Thus the sequence $\{A_{2N+1}\}$ is increasing.

We have two monotone sequences. We would like to show that they are bounded so that we can apply the Monotone Sequence Theorem. Observe that, for all N,

$$A_{2N} = A_{2N-1} + \frac{1}{2N+1} > A_{2N-1} > A_1,$$

that is, the decreasing sequence $\{A_{2N}\}$ is bounded below by A_1. It is therefore convergent. Call its limit L. Similarly, for all N,

$$A_{2N+1} = A_{2N} - \frac{1}{2N+2} < A_{2N} < A_0,$$

that is, the increasing sequence $\{A_{2N+1}\}$ is bounded above by A_0. It is therefore convergent. Call its limit K. We claim that $L = K$. Indeed,

$$\begin{aligned} K - L &= \lim_{N\to\infty} A_{2N+1} - \lim_{N\to\infty} A_{2N} \\ &= \lim_{N\to\infty} (A_{2N+1} - A_{2N}) \\ &= \lim_{N\to\infty} -\frac{1}{2N+2} \\ &= 0. \end{aligned}$$

Call this common limit L. We claim $\{A_N\}$ converges to L. Let $\varepsilon > 0$ be given. Because $\{A_{2J}\}$ and $\{A_{2J+1}\}$ converge to L, there exists M_1 such that, for all $J \geq M_1$, $|A_{2J} - L| < \varepsilon$, and there exists M_2 such that, for all $J \geq M_2$, $|A_{2J+1} - L| < \varepsilon$. Let $M = \max\{M_1, M_2\}$ and suppose $N \geq 2M + 1$. If N is even, $N = 2J$ for some J, and this J satisfies $2J \geq 2M + 1$, hence $J \geq M$. If N is odd, then $N = 2J + 1$ for some J and this J satisfies $2J + 1 \geq 2M + 1$, and hence again $J \geq M$. In either case, we conclude $|A_N - L| < \varepsilon$. We have finally proved that $\sum_{n=0}^{\infty} \frac{(-1)^n}{n+1}$ is convergent.

The series (14.1) is called the *alternating harmonic series.* In general, any series in which the sign of the terms depends on the parity of the index is called an *alternating series.* We will prove a general theorem about alternating series in the next section. For now, we are still in the exploration phase.

Exercise 14.1. Consider the alternating series $\sum_{n=0}^{\infty} \frac{(-1)^n}{\sqrt{n+1}}$.

(a) Is this series absolutely convergent?

(b) Plot the first 100 partial sums of the alternating series and make some conjectures about the sequence $\{A_N\}$ of partial sums and the subsequences $\{A_{2N}\}$ and $\{A_{2N+1}\}$.

(c) Prove that $\sum_{n=0}^{\infty} \frac{(-1)^n}{\sqrt{n+1}}$ converges.

Exercise 14.2. Consider the alternating series $\sum_{n=0}^{\infty}(-1)^n \frac{n}{2n+1}$. Plot its first 50 partial sums. Then prove that the series diverges.

14.1.2 Summation by Parts and the Alternating Series Test

We have now seen several examples of convergent and divergent alternating series. Of course, in all the convergent examples, the terms of the series have limit 0. We know that for arbitrary series, the condition that the terms have limit 0 is necessary but not sufficient for the convergence of the series. We might hope that for alternating series, the condition is sufficient. Unfortunately it is not. (See Problem 3.) We can, however, add one additional simple hypothesis and we obtain a useful theorem.

Theorem 14.1 (Alternating Series Test). *Let $\{a_n\}$ be a sequence of non-negative real numbers. If*

(i) $\{a_n\}$ *is non-increasing, and*

(ii) $\lim_{n\to\infty} a_n = 0$,

then the alternating series $\sum(-1)^n a_n$ *converges.*

Several proofs are possible. Our first proof is a general version of the argument we used to prove the convergence of the alternating harmonic series.

Proof. As usual, let $\{A_N\}$ be the sequence of partial sums for the series. We consider the two subsequences $\{A_{2N}\}$ and $\{A_{2N+1}\}$. For $N \geq 1$,

$$A_{2N} = A_{2N-2} + (-a_{2N-1} + a_{2N}).$$

Because $\{a_n\}$ is non-increasing, $a_{2N-1} \geq a_{2N}$. Therefore $-a_{2N-1} + a_{2N} \leq 0$, and so $A_{2N} \leq A_{2N-2}$. We conclude that $\{A_{2N}\}$ is non-increasing. Similarly, for all $N \geq 1$,

$$A_{2N+1} = A_{2N-1} + (a_{2N} - a_{2N+1}).$$

Because $\{a_n\}$ is non-increasing, $a_{2N} - a_{2N+1} \geq 0$, and so $A_{2N+1} \geq A_{2N-1}$. We conclude that $\{A_{2N+1}\}$ is non-decreasing. Furthermore, for all $N \geq 1$,

$$A_{2N} = A_{2N-1} + a_{2N} \geq A_{2N+1} \geq A_1,$$

and

$$A_{2N+1} = A_{2N} - a_{2N+1} \leq A_{2N} \leq A_0.$$

We may thus apply the Monotone Sequence Theorem to conclude that $\{A_{2N}\}$ converges to some real number L and $\{A_{2N+1}\}$ converges to some real number K. In fact $L = K$, for

$$\begin{aligned} K - L &= \lim_{N\to\infty} A_{2N+1} - \lim_{N\to\infty} A_{2N} \\ &= \lim_{N\to\infty} (A_{2N+1} - A_{2N}) \\ &= \lim_{N\to\infty} a_{2N+1} \\ &= 0. \end{aligned}$$

Call the common limit L. Now the same argument given at the end of the proof of the convergence of the alternating harmonic series proves that $\{A_N\}$ itself converges to L. The proof is complete. □

One great advantage of this proof of the alternating series test is that it allows us to estimate the error we make when we use a partial sum of such a convergent alternating series to approximate the full series. We state this estimate as a corollary.

Corollary 14.1. *Let $\{a_n\}$ satisfy the hypotheses of Theorem 14.1. If we use the N-th partial sum A_N to approximate the sum L of the convergent alternating series $\sum(-1)^n a_n$, then $|A_N - L| \leq a_{N+1}$.*

Proof. We consider two cases, depending on the parity of N. Suppose N is even, so that $N = 2J$ for some J. Because $A_{2J+1} \leq L \leq A_{2J}$,

$$\begin{aligned} |A_N - L| &= A_{2J} - L \\ &\leq A_{2J} - A_{2J+1} \\ &= -(-1)^{2J+1} a_{2J+1} \\ &= a_{N+1}. \end{aligned}$$

The proof for N odd is similar and is omitted. □

Example 14.1. Return to the alternating harmonic series (14.1). Suppose we want to estimate its value with an error less than 0.0001. By Corollary 14.1, if we find N so that $a_{N+1} \leq 0.0001$, then for that N, A_N will be within 0.0001 of the true sum. We find that $N = 9998$ and $A_{9998} = 0.693197$. (It is possible to show that, in fact, the alternating harmonic series converges to $\ln(2)$, but the proof is beyond the scope of this text.)

Example 14.2. Consider $\sum_{n=0}^{\infty} \frac{(-1)^n}{\sqrt[3]{n+1}}$. Prove that this series converges and estimate its value to within 0.01.

Our second proof of the alternating series test will be an application of a theorem that is very useful in its own right.

Theorem 14.2 (Summation by Parts). *Let $\{a_n\}$ and $\{b_n\}$ be sequences of real numbers. Let $B_N = \sum_{n=1}^{N} b_n$. Then*

$$\sum_{n=1}^{N} a_n b_n = a_N B_N - \sum_{n=1}^{N-1} (a_{n+1} - a_n) B_n. \tag{14.2}$$

Proof. The proof is by induction on N. When $N = 1$, the left-hand side of (14.2) is $a_1 b_1$. On the right-hand side, the sum $\sum_{n=1}^{0}(a_{n+1} - a_n)B_n$ is empty, hence equal to 0, and $B_1 = b_1$. The right-hand side thus also equals $a_1 b_1$, and so the result holds for $N = 1$.

Now suppose the result holds for $K \geq 1$ and consider $K + 1$. Then

$$\begin{aligned} \sum_{n=1}^{K+1} a_n b_n &= a_{K+1} b_{K+1} + \sum_{n=1}^{K} a_n b_n \\ &= a_{K+1} b_{K+1} + a_K B_K - \sum_{n=1}^{K-1} (a_{n+1} - a_n) B_n, \end{aligned} \tag{14.3}$$

by the inductive hypothesis. We now replace b_{K+1} with $B_{K+1} - B_K$ and do additional algebraic manipulations to obtain

$$\begin{aligned}
(14.3) \quad &= a_{K+1}(B_{K+1} - B_K) + a_K B_K - \sum_{n=1}^{K-1}(a_{n+1} - a_n)B_n \\
&= a_{K+1}B_{K+1} - (a_{K+1} - a_K)B_K - \sum_{n=1}^{K-1}(a_{n+1} - a_n)B_n \\
&= a_{K+1}B_{K+1} - \sum_{n=1}^{K}(a_{n+1} - a_n)B_n.
\end{aligned}$$

Thus the result holds for $K+1$ and, by the principle of mathematical induction, it holds for all N. □

Summation by parts can be used to evaluate some finite series. It allows us to trade in a series $\sum_{n=1}^{N} a_n b_n$ we can't do for a series $\sum_{n=1}^{N-1}(a_{n+1} - a_n)B_n$ that we might be able to do. The name and form of the theorem may remind you of a result you know from calculus, namely integration by parts. Suppose you have an integral of a product of functions a and b that you can not do. Integration by parts says that if you can find an antiderivative B for b, then you can trade in the hard integral for the hopefully better integral of a' times B. In symbols,

$$\int_c^d a(x)b(x)\,dx = (a(d)B(d) - a(c)B(c)) - \int_c^d a'(x)B(x)\,dx.$$

For example, if we want to do $\int_0^1 x\cos x\,dx$, we set $a(x) = x$ and $b(x) = \cos x$. Then $a'(x) = 1$, $B(x) = \sin x$, and

$$\int_0^1 x\cos x\,dx = (1 \cdot \sin 1 - 0 \cdot \sin 0) - \int_0^1 \cos x\,dx.$$

This new integral is easy to do. Summation by parts works in an analogous manner.

Exercise 14.3. Find and prove a formula for $\sum_{n=1}^{N} n2^n$.

Summation by parts can also be used to rewrite the partial sums of an infinite series and thus may lead to a proof of convergence or divergence for that series. We return to our favorite series of the chapter to illustrate.

Example 14.3. Consider the N-th partial sum $A_N = \sum_{n=0}^{N} \frac{(-1)^n}{n+1}$ for the alternating harmonic series. We will apply summation by parts with $a_n = \frac{1}{n+1}$ and $b_n = (-1)^n$. The summation by parts formula requires us to understand $B_N = \sum_{n=0}^{N} b_n$. We find that

$$B_N = \sum_{n=0}^{N}(-1)^n = \begin{cases} 1 & \text{if } N \text{ is even} \\ 0 & \text{if } N \text{ is odd} \end{cases}.$$

$\{B_N\}$ is not a convergent sequence, but it is bounded.

We now rewrite the partial sums using summation by parts:

$$\begin{aligned} A_N &= \sum_{n=0}^{N} \frac{1}{n+1} \cdot (-1)^n \\ &= \frac{1}{N+1} B_N - \sum_{n=0}^{N-1} \left(\frac{1}{n+2} - \frac{1}{n+1} \right) B_n \\ &= \frac{1}{N+1} B_N + \sum_{n=0}^{N-1} \frac{1}{(n+1)(n+2)} B_n. \end{aligned}$$

Now we consider the limit as $N \to \infty$. Because $\lim_{N\to\infty} \frac{1}{N+1} = 0$ and $\{B_N\}$ is bounded, $\lim_{N\to\infty} \frac{1}{N+1} B_N = 0$. Next consider $\sum_{n=0}^{N-1} \frac{1}{(n+1)(n+2)} B_n$. The terms of this sum are non-negative and bounded above by the terms of the series $\sum_{n=0}^{\infty} \frac{1}{(n+1)^2}$. Because the latter is a convergent p-series (with $p = 2$), by the comparison test, $\sum_{n=0}^{\infty} \frac{1}{(n+1)(n+2)} B_n$ is convergent. Thus

$$\lim_{N\to\infty} \sum_{n=0}^{N-1} \frac{1}{(n+1)(n+2)} B_n$$

exists. We conclude that $\lim_{N\to\infty} A_N$ exists, completing our proof that the alternating harmonic series is convergent.

This example gives a nice template we can follow in our general proof of the alternating series test using summation by parts. It also draws our attention to several lemmas we need in such arguments. We state these lemmas and leave the proofs of the first two as exercises because they are very easy. We intentionally use notation that is consistent with the notation that will appear in the proof to follow.

Lemma 14.1. *Let $\{a_N\}$ and $\{B_N\}$ be sequences of real numbers. If $\lim_{N\to\infty} a_N = 0$ and if $\{B_N\}$ is bounded, then $\lim_{N\to\infty} a_N B_N = 0$.*

Lemma 14.2. *Let $\{a_n\}$ and $\{B_n\}$ be sequences of real numbers. If $\sum_{n=0}^{\infty} |a_{n+1} - a_n|$ converges and $\{B_n\}$ is bounded, then $\sum_{n=0}^{\infty} (a_{n+1} - a_n) B_n$ converges absolutely.*

Lemma 14.3. *Let $\{a_n\}$ be a sequence of real numbers. If $\{a_n\}$ is non-increasing and $\lim_{n\to\infty} a_n = 0$, then $\sum_{n=0}^{\infty} |a_{n+1} - a_n|$ converges.*

Proof. Fix N. We must show that $\lim_{N\to\infty} \sum_{n=0}^{N} |a_{n+1} - a_n|$ exists. Because $\{a_n\}$ is non-increasing, for each n, $a_{n+1} \leq a_n$, that is, $a_{n+1} - a_n \leq 0$. Thus for all n, $|a_{n+1} - a_n| = a_n - a_{n+1}$. Then

$$\begin{aligned}
&\sum_{n=0}^{N} |a_{n+1} - a_n| \\
&= \sum_{n=0}^{N} (a_n - a_{n+1}) \\
&= (a_0 - a_1) + (a_1 - a_2) + \ldots + (a_{N-1} - a_N) + (a_N - a_{N+1}).
\end{aligned}$$

This sum is telescoping, so it simplifies to $a_0 - a_{N+1}$. Because $\lim_{N\to\infty} a_{N+1} = 0$,

$$\lim_{N\to\infty} \sum_{n=0}^{N} |a_{n+1} - a_n| = \lim_{N\to\infty} (a_0 - a_{N+1}) = a_0.$$

The conclusion follows. □

Exercise 14.4. Prove Lemmas 14.1 and 14.2.

We may now give our second proof of the alternating series test.

Proof. Consider $A_N = \sum_{n=0}^{N} (-1)^n a_n$. Apply summation by parts with $b_n = (-1)^n$. As we saw in Example 14.3, $\{B_n\}$ is a bounded sequence that is equal to 0 or 1 for every n. By Theorem 14.2,

$$A_N = a_N B_N - \sum_{n=0}^{N-1} (a_{n+1} - a_n) B_n. \tag{14.4}$$

We wish to take the limit of (14.4) as $N \to \infty$. By Lemma 14.1, the first term on the right has limit 0. By Lemma 14.3, $\sum_{n=0}^{\infty} |a_{n+1} - a_n|$ converges, and by Lemma 14.2, $\sum_{n=0}^{\infty} (a_{n+1} - a_n) B_n$ converges absolutely. Thus the limit of the second term on the right of (14.4) also exists. We conclude that $\lim_{N\to\infty} A_N$ exists. Hence the corresponding series is convergent. □

14.1.3 Basic facts about conditionally convergent series

Before we go on to discuss the subtleties of conditionally convergent series, we establish a number of their elementary properties. In particular, it is intuitively clear that a conditionally convergent series must have infinitely many positive and negative terms and that the corresponding series of positive terms and series of negative terms should both be divergent. We will prove these statements.

First, we establish some notation. Let $\{a_n\}$ be a sequence of real numbers. For each n, define sequences a^+ and a^- by the rules

$$a_n^+ = \begin{cases} a_n & \text{if } a_n \geq 0 \\ 0 & \text{otherwise} \end{cases}$$

and

$$a_n^- = \begin{cases} a_n & \text{if } a_n < 0 \\ 0 & \text{otherwise} \end{cases}.$$

The sequences a^+ and a^- are *not* subsequences of a. The sequence a^+ agrees with a for every n for which a_n is non-negative, and it is equal to 0 for any n for which a_n is negative. For example, for the alternating harmonic series,

$$\begin{aligned} a^+ &= \left\{1, 0, \frac{1}{3}, 0, \frac{1}{5}, 0, \ldots\right\} \\ a^- &= \left\{0, -\frac{1}{2}, 0, -\frac{1}{4}, 0, \ldots\right\}. \end{aligned}$$

As usual, let A be the sequence of partial sums for $\sum a_n$, and let A^+ and A^- be the sequences of partial sums for $\sum a_n^+$ and $\sum a_n^-$, respectively. Although a^+ is not a subsequence of a, $\lim_{N\to\infty} A_N^+$ is the value of the series consisting of all non-negative terms of a if this series converges or it is the value $+\infty$ if this series diverges. A similar statement can be made about $\lim_{N\to\infty} A_N^-$. Furthermore, for all N, $A_N = A_N^+ + A_N^-$. Also, for all N,

$$\sum_{n=0}^{N} |a_n| = \sum_{n=0}^{N} (a_n^+ - a_n^-) = A_N^+ - A_N^-.$$

We are ready to state our propositions.

Proposition 14.1. *Let $\{a_n\}$ be a sequence of real numbers, and suppose $\sum a_n$ is conditionally convergent. Then the sequences a^+ and a^- both have infinitely many non-zero terms.*

Proof. At least one of a^+ and a^- must have infinitely many non-zero terms, for otherwise $\sum |a_n|$ would be convergent. Suppose, for a contradiction, that one of the sequences has only finitely many non-zero terms. Suppose for definiteness that it is a^-. Let $K = \max\{n : a_n^- \neq 0\}$ (or 0 if this set is empty). Then for all $N \geq K$, $A_N^- = A_K^-$. For convenience, call this sum M. Now, if $N \geq K$,

$$\begin{aligned} \sum_{n=0}^{N} |a_n| &= A_N^+ - A_N^- \\ &= A_N^+ + A_N^- - 2A_N^- \\ &= A_N - 2M. \end{aligned}$$

Because $\sum a_n$ converges, $\lim_{N\to\infty} A_N$ exists, and hence $\lim_{N\to\infty}(A_N - 2M)$ exists. We conclude that $\sum |a_n|$ converges. This conclusion contradicts the hypothesis that $\sum a_n$ is only conditionally convergent, so we conclude that in fact a^- and a^+ both have infinitely many non-zero terms. □

The next proposition confirms our intuition that conditionally convergent series converge because of cancellation between the positive and negative terms, for it says that either set of terms alone gives rise to a divergent series. The proof is very similar to the proof of Proposition 14.1 and is left as an exercise.

Proposition 14.2. *Let $\{a_n\}$ be a sequence of real numbers, and suppose $\sum a_n$ is conditionally convergent. Then $\sum a_n^+$ and $\sum a_n^-$ are both divergent.*

Exercise 14.5. Prove Proposition 14.2.

14.2 Rearrangements

Suppose you were asked to find the sum

$$7 + 11 + 3 + 4 + 9 + 6.$$

A quick way to do it would be to rearrange and group the numbers like so:

$$(7 + 3) + (11 + 9) + (6 + 4) = 10 + 20 + 10 = 40.$$

The commutative law for addition allows us to change the order of a sum of two real numbers, and the associative law allows us to group two numbers together in a sum of three. Using mathematical induction, we can extend these results to show that we can rearrange and group any finite sum without changing its value. However, mathematical induction can not be used to extend this result to the infinite case. The question remains: Can we rearrange or group the terms of an infinite series to make the sum easier to evaluate? We considered this question briefly in Chapter 10 when we studied decimal representations of real numbers. In that setting, all of the terms of our series were *non-negative*. We said then that rearranging and grouping terms of non-negative series does not change the value of the sum, but we indicated that series with positive and negative terms are trickier. In this section we address these questions.

14.2.1 Rearrangements and non-negative series

The intuitive definition of a rearrangement is easy; $\sum b_n$ is a rearrangement of $\sum a_n$ if the series $\sum b_n$ includes all the terms of $\sum a_n$, but possibly in a different order. Thus there exists $n(0)$ such that $b_0 = a_{n(0)}$, there exists $n(1)$ different from $n(0)$ so that $b_1 = a_{n(1)}$, and, in general, for $k \geq 1$, there exists $n(k)$ different from $n(0), \ldots, n(k-1)$ such that $b_k = a_{n(k)}$. What properties does the function $k \mapsto n(k)$ have? It is certainly a function from $\mathbb{N}_0$ to $\mathbb{N}_0$. It must be surjective because every term in the series $\sum a_n$ must appear in the rearrangement, and it must be injective because every term in $\sum a_n$ appears only once in the rearrangement. We summarize succinctly in a definition.

Definition 14.3. Let a and b be sequences of real numbers. Then $\sum b_k$ is a **rearrangement** of $\sum a_n$ if there exists a bijection $n\colon \mathbb{N}_0 \to \mathbb{N}_0$ such that $b = a \circ n$ (i.e., $b_k = a_{n(k)}$ for all k). We call $k \mapsto n(k)$ the rearrangement function.

We are now ready to prove a theorem about rearrangements of non-negative series.

Theorem 14.3. *Let $\{a_n\}$ be a sequence of non-negative real numbers. If $\sum a_n$ converges with sum L, then every rearrangement of $\sum a_n$ converges with sum L.*

Proof. Let $A_N = \sum_{n=0}^{N} a_n$ denote the N-th partial sum of the original series, and let $B_K = \sum_{k=0}^{K} b_k$ denote the K-th partial sum of the rearrangement. We must show that $\lim_{K\to\infty} B_K = L$. Because the a_n are non-negative, $\{A_N\}$ and $\{B_K\}$ are non-decreasing sequences, and $A_N \leq L$ for all N. It is also true that $B_K \leq L$ for all K. Indeed, given K, we can find $N(K)$ large enough so that every term contributing to the partial sum B_K is one of the terms contributing to $A_{N(K)}$ and thus $B_K \leq A_{N(K)} \leq L$.

Let $\varepsilon > 0$ be given. Because $\{A_N\}$ converges to L, there exists M such that, for all $N \geq M$, $|A_N - L| < \varepsilon$. Fix such an N and let J be such that B_J includes all of the terms $a_0, \ldots, a_N$. Thus $B_J \geq A_N$. Then for $K \geq J$,

$$\begin{aligned} |B_K - L| &= L - B_K \\ &\leq L - B_J \\ &\leq L - A_N \\ &< \varepsilon, \end{aligned}$$

as desired. □

Exercise 14.6. Modify the proof of Theorem 14.3 to show that, if $\{a_n\}$ is a sequence of real numbers and if $\sum a_n$ converges absolutely, with sum L, then any rearrangement converges, with sum L.

14.2.2 Using Python to explore the alternating harmonic series

Conditionally convergent series are a different animal altogether. When a series converges conditionally, the convergence occurs because of cancellation; positive and negative terms combine in just the right way to produce the limit. If we change the order of the terms, we might destroy this cancellation. In the next subsection, we will prove an amazing result that says a conditionally convergent series can be rearranged to produce anything we want. For any real number L, there are rearrangements that converge to L. There are also rearrangements for which the partial sums tend to $+\infty$ or $-\infty$. Because this result is so startling, we first experiment with rearrangements of the alternating harmonic series.

We have seen that the alternating harmonic series converges to a number close to 0.6932. As a modest first goal, let's produce a rearrangement that converges to 1.1. The idea is rather simple. Begin by adding the positive terms, in order, until the sum first reaches or exceeds the target of 1.1. Then add negative terms until the sum drops below 1.1. Then add positive terms again until the sum again reaches or exceeds 1.1. Continue this process.

We can begin to carry out this process by hand, constructing the rearranged sequence b. Let S be the sequence of partial sums of the rearrangement. Because the first positive term of the sequence is 1, we set $b_0 = a_0 = 1$. Then $S_0 = 1$ as well. Because S_0 is not greater than or equal to our target of 1.1, we add the next positive term, taking $b_1 = a_2 = \frac{1}{3}$. Now

$$S_1 = 1 + \frac{1}{3} = \frac{4}{3} \geq 1.1,$$

so we begin adding negative terms, setting $b_2 = a_1 = -\frac{1}{2}$. Now

$$S_2 = S_1 - \frac{1}{2} = \frac{4}{3} - \frac{1}{2} = \frac{5}{6} < 1.1,$$

so we begin adding positive terms again, setting $b_3 = a_4 = \frac{1}{5}$. Because

$$S_3 = S_2 + \frac{1}{5} = \frac{5}{6} + \frac{1}{5} = \frac{31}{30} < 1.1,$$

we must add another positive term. We set $b_4 = a_6 = \frac{1}{7}$. Now

$$S_4 = \frac{31}{30} + \frac{1}{7} = \frac{247}{210} > 1.1.$$

We are again ready to add negative terms. We take $b_5 = a_3 = -\frac{1}{4}$. Our rearranged sequence up to this point is

$$1, \frac{1}{3}, -\frac{1}{2}, \frac{1}{5}, \frac{1}{7}, -\frac{1}{4}, \ldots.$$

Constructing a rearrangement by hand is very tedious and time-consuming. But we can write a short program that will allow us to obtain the first terms in a rearrangement quickly. Let's write code to generate the first 50 terms in the rearrangement of the alternating harmonic series converging to 1.1. As above, we let b denote the rearranged sequence and we let S denote the sequence of partial sums of the rearrangement. We need to keep track of the numbers of positive and negative terms we use. We let p be the counter for the positive terms and we let q be the counter for the negative terms. Of course these are initially set to 0. We also define a variable *current* to hold the running total of all the terms in the rearrangement at a given step. Of course we initially set *current* $= 0$.

We now go through a loop 50 times to generate 50 terms of the rearrangement. Each time through, we compare the current running total to the target. If *current* is less than the target, we add the next positive term, which is $a_{2p} = \frac{1}{2p+1}$. This term is appended to the rearranged sequence b, and the current total is appended to the sequence S of partial sums of the rearrangement. Because we added a positive term, we increment the positive counter. If, on the other hand, *current* meets or exceeds the target, we add the next negative term of the sequence, which is $a_{2q+1} = -\frac{1}{2q+2}$. We append this term to b, add the current total to S, and increment the negative counter. Here's the code:

```
target=1.1

b=[] #holds terms of rearrangement
S=[] #holds partial sums of rearrangement

p=0 #positive counter
q=0 #negative counter
current=0.0

for i in range(0,50):
    if current<target: #add the next positive term
        current=current+1.0/(2*p+1)
        b.append(1.0/(2*p+1))
        S.append(current)
        p=p+1
    else: #add the next negative term
        current=current-1.0/(2*q+2)
        b.append(-1.0/(2*q+2))
        S.append(current)
        q=q+1

plot(b,"v") #use different markers for the two sequences
plot(S,"^")
show()
```

Figure 14.2 shows the plot. Triangles pointing down indicate terms in b and triangles pointing up indicate terms in S.

Exercise 14.7. Modify the code to find the first 100 terms in the rearrangements of the alternating harmonic series converging to 0, 1, 2, and 3. In each case, in addition to the plots, have the program give as output the numbers of positive and negative terms used in the rearrangements.

14.2.3 A general theorem

One of the main themes of this book has been that an algorithm to approximate an analytic object and a constructive proof of its existence go hand-in-hand; if we can write an algorithm to do it, we can use the idea of the algorithm to construct a proof, and vice versa. We use the idea of the algorithm from the previous subsection to prove a general theorem about rearrangements.

Theorem 14.4 (Riemann Rearrangement Theorem). *Let $a = \{a_n\}$ be a sequence of real numbers indexed by $\mathbb{N}$, and suppose $\sum a_n$ is conditionally convergent. Let L be an extended real number (i.e., either $L \in \mathbb{R}$ or $L = \pm\infty$). Then there is a rearrangement $\sum b_n$ of $\sum a_n$ whose partial sums have limit L.*

Proof. The proof is very notation-intensive, so it is important to keep in mind the basic idea of the algorithm. We treat the case in which the target sum L is finite, leaving the cases in which $L = \pm\infty$ to the exercises.

Figure 14.2: The rearranged sequence b and the partial sums of the rearrangement.

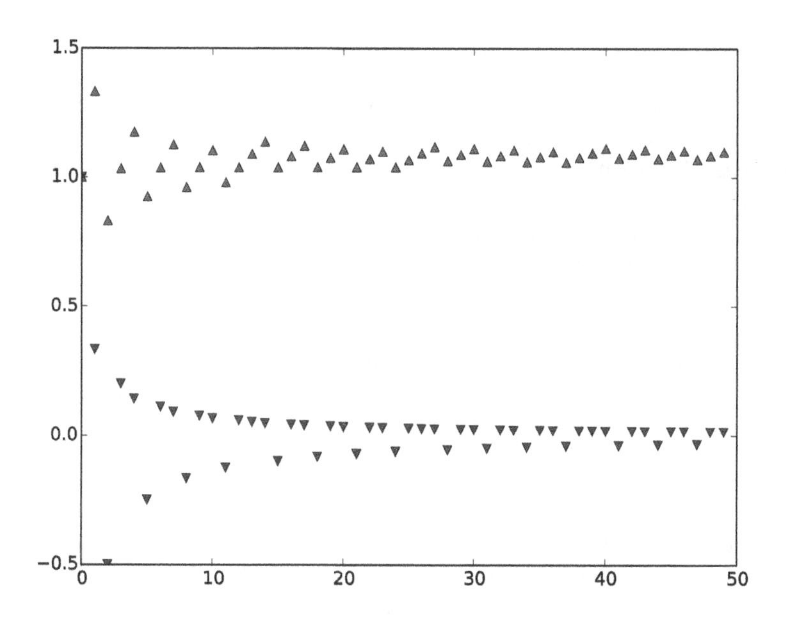

Let c denote the subsequence of a consisting of the non-negative terms of a, in order, and let d be the subsequence of a consisting of the negative terms of a, in order. By Proposition 14.1, both c and d have infinitely many non-zero terms. Our goal is to construct a sequence b so that $\sum b_n$ is a rearrangement of $\sum a_n$ converging to L. We use capital letters to denote the sequence of partial sums of a series, so that A is the sequence of partial sums of $\sum a_n$, B is the sequence of partial sums of $\sum b_n$, and so on. By Proposition 14.2, both series $\sum c_n$ and $\sum d_n$ are divergent. Because all terms in each series are of constant sign, $\lim_{N\to\infty} C_N = \infty$ and $\lim_{N\to\infty} D_N = -\infty$. As a consequence, at every point in our construction, we will only need finitely many of the remaining terms to achieve a sum as large as we need. We use this fact without further comment throughout the remainder of the proof.

We now select terms for the rearranged sequence b. First we add enough non-negative terms so that their sum meets or exceeds L. That is, let p_1 be the smallest element of $\mathbb{N}$ for which $C_{p_1} = \sum_{n=1}^{p_1} c_n \geq L$. Then we let q_1 be the smallest element of $\mathbb{N}$ for which $C_{p_1} + D_{q_1} < L$. Thus the sequence b begins with

$$c_1, \ldots, c_{p_1}, d_1, \ldots, d_{q_1}.$$

These are the terms b_1 up to $b_{p_1+q_1}$ of b. One of these b_j must equal a_1. Note that the corresponding partial sum of the rearrangement is $B_{p_1+q_1} = C_{p_1} + D_{q_1}$.

Proceed inductively; having obtained p_{k-1} and q_{k-1} for $k \geq 2$, let p_k be the smallest natural number greater than p_{k-1} for which $C_{p_k} + D_{q_{k-1}} \geq L$, and let q_k be the smallest natural number greater than q_{k-1} for which $C_{p_k} + D_{q_k} < L$. At this k-th step, we append

$$c_{p_{k-1}+1}, \ldots, c_{p_k}, d_{q_{k-1}+1}, \ldots, d_{q_k}$$

to the end of the portion of the sequence b previously obtained. At the end of this step, our b sequence includes terms $b_1, \ldots, b_{p_k+q_k}$. Also, because we have chosen at least k positive terms and at least k negative terms, these b_j must include $a_1, \ldots, a_k$.

We must show that the partial sums B_N of the rearrangement have limit L. We will obtain estimates on B_N first for N of a special form and then for general N.

Case 1: $N = p_k + q_k - 1$. For N of this form, we are in the process of adding negative terms to our rearrangement. In fact, by the definition of q_k, $B_{p_k+q_k-1} = C_{p_k} + D_{q_k-1} \geq L$, but $B_{p_k+q_k-1} + d_{q_k} < L$. Combining these inequalities (recalling that $d_{q_k} < 0$) gives

$$L \leq B_{p_k+q_k-1} < L + |d_{q_k}|. \tag{14.5}$$

In words, the current sum is greater than or equal to L, but adding one more negative term will be enough to make the sum drop below L.

Case 2: $N = p_k + q_k$. We obtain this inequality by adding d_{q_k} to all three expressions in (14.5), noting that $B_{p_k+q_k-1} + d_{q_k} = B_{p_k+q_k}$. We obtain

$$L - |d_{q_k}| \leq B_{p_k+q_k} < L. \tag{14.6}$$

Case 3: $N = p_k + q_{k-1} - 1$. For N of this form, we are in the process of adding positive terms to our rearrangement. In fact, by the definition of p_k, $B_{p_k+q_{k-1}-1} = C_{p_k-1} + D_{q_{k-1}} < L$, but $B_{p_k+q_{k-1}-1} + c_{p_k} \geq L$. Combining these inequalities gives

$$L - c_{p_k} \leq B_{p_k+q_{k-1}-1} < L. \tag{14.7}$$

In words, the current sum is less than L, but adding one more positive term will be enough to make it greater than or equal to L.

Case 4: $N = p_k + q_{k-1}$. We need only add c_{p_k} to all terms of the inequality (14.7) to obtain

$$L \leq B_{p_k+q_{k-1}} < L + c_{p_k}. \tag{14.8}$$

Now we consider arbitrary N. Assume for simplicity that $N \geq p_1 + q_1$. Because the numbers $p_k + q_k$ tend to infinity, there exists k such that

$$p_k + q_k \leq N < p_{k+1} + q_{k+1}.$$

Suppose first that $p_k + q_k < N \leq p_{k+1} + q_k$. Then we are in the process of adding positive terms to the sum, and

$$L - |d_{q_k}| \leq B_{p_k+q_k} \leq B_N \leq B_{p_{k+1}+q_k} < L + c_{p_{k+1}}.$$

If, on the other hand, $p_{k+1} + q_k < N < p_{k+1} + q_{k+1}$, we are in the process of adding negative terms to the sum, and

$$L - |d_{q_{k+1}}| \leq B_{p_{k+1}+q_{k+1}} \leq B_N \leq B_{p_{k+1}+q_k} \leq L + c_{p_{k+1}}.$$

To summarize, if $p_k + q_k \leq N < p_{k+1} + q_{k+1}$, then

$$|B_N - L| \leq \max\{|d_{q_k}|, |d_{q_{k+1}}|, c_{p_{k+1}}\}. \tag{14.9}$$

Let $\varepsilon > 0$ be given. Because $\lim_{n\to\infty} a_n = 0$, there exists K such that, if $n \geq K$, $|a_n| < \varepsilon$. All of $a_1, \ldots, a_{K-1}$ are guaranteed to be in $b_1, \ldots, b_{p_{K-1}+q_{K-1}}$. Take $N \geq p_K + q_K$. Then for some $k \geq K$, (14.9) holds. Because all of d_{q_k}, $d_{q_{k+1}}$, and $c_{p_{k+1}}$ are terms of the sequence a for indices bigger than N,

$$|B_N - L| \leq \max\{|d_{q_k}|, |d_{q_{k+1}}|, c_{p_{k+1}}\} < \varepsilon.$$

Therefore $\lim_{N\to\infty} B_N = L$. □

Exercise 14.8. Prove Theorem 14.4 for $L = +\infty$. That is, prove that if $\sum a_n$ is conditionally convergent, there is a rearrangement whose partial sums tend to $+\infty$.

14.3 Problems

1. Give an example of a sequence $\{a_n\}$ such that $\sum_{n=0}^{\infty} a_n$ converges but $\sum_{n=0}^{\infty} a_n^2$ diverges.

2. Give an example of a sequence $\{a_n\}$ such that $\sum_{n=0}^{\infty} a_n$ converges but $\sum_{n=0}^{\infty} a_n^3$ diverges.

3. Consider the alternating series $\sum_{n=0}^{\infty}(-1)^n a_n$ with

$$a_n = \begin{cases} \frac{1}{n+1} & \text{if } n \text{ is odd} \\ \frac{1}{\sqrt{n+1}} & \text{if } n \text{ is even.} \end{cases}$$

   (a) Plot the first 50, the first 500, and the first 5000 partial sums of the series. Make a conjecture about the convergence of the series.

   (b) Consider the subsequences $\{A_{2N}\}$ and $\{A_{2N+1}\}$ of the sequence of partial sums. How are they similar to the corresponding sequences for the alternating harmonic series? How are they different?

   (c) Show that there is a constant c so that $A_{2N} - A_{2N-2} \geq \frac{c}{\sqrt{N}}$.

   (d) Prove that $\{A_{2N}\}$ diverges. Conclude that the original series diverges.

4. Find all real numbers x for which each of the following series converges.

   (a) $\displaystyle\sum_{n=1}^{\infty} \frac{x^n}{n}$.

(b) $\displaystyle\sum_{n=1}^{\infty} \frac{x^n}{2^n \sqrt{n}}$.

(c) $\displaystyle\sum_{n=1}^{\infty} \frac{(x-2)^n}{3^n n}$.

5. Give an example of each of the following or explain why no such series exists.

   (a) An alternating series that is absolutely convergent.

   (b) An alternating series with unbounded partial sums.

   (c) An absolutely convergent series with a divergent rearrangement.

6. There are many ways to define the exponential and trigonometric functions. One way is using series. For $x \in \mathbb{R}$ or $\mathbb{C}$ define

$$\begin{aligned} e^x &= \sum_{n=0}^{\infty} \frac{x^n}{n!} \\ \sin x &= \sum_{n=0}^{\infty} \frac{(-1)^n}{(2n+1)!} x^{2n+1} \\ \cos x &= \sum_{n=0}^{\infty} \frac{(-1)^n}{(2n)!} x^{2n}. \end{aligned}$$

   (a) Find all x for which each of the above series converges.

   (b) One can prove that each of these functions is differentiable at every point of its domain and that the derivative is obtained by differentiating each series term by term. Assume this result and find the derivative of each function.

   (c) Show that for all x, $\sin^2 x + \cos^2 x = 1$. (Hint: Find the derivative of $f(x) = \sin^2 x + \cos^2 x$ and use Problem 9 in Chapter 13.)

7. Refer to the previous problem.

   (a) Let $t \in \mathbb{R}$. Find the real and imaginary parts of e^{it}.

   (b) Fix $x \in \mathbb{R}$. Find $\sum_{n=1}^{N} e^{inx}$.

   (c) (Challenging) For which, if any, $x \in \mathbb{R}$ does $\sum_{n=1}^{\infty} \frac{\sin nx}{n}$ converge? (Hint: Write $\sin nx$ in terms of e^{inx} and use summation by parts.)

Programming Project. Write code to generate the first 100 terms in a rearrangement of the alternating harmonic series tending to ∞. Plot the rearranged sequence and the partial sums of the rearrangement.

Appendix A

A Very Short Course on Python

This appendix contains everything you need to understand the Python examples in this book and to write your own simple programs as requested in the exercises. We recommend that you work through Sections 1, 2, and 3 before you read Chapter 2 in the text, for in Chapter 2 we write a number of programs to help us explore sequences. The material in Sections 4 and 5 is not used heavily until Chapter 5.

Many books and online resources give an encyclopedic treatment of Python. Our treatment is purposely *not* encyclopedic. We aim instead to give a very short introduction to the major ideas. This treatment is intended to be readable to those with absolutely no prior programming experience. Sometimes students are intimidated by the thought of using computers to explore mathematical problems, thinking that they must be fluent in a programming language before they can start using it. But we argue that learning a programming language is like learning any other language; it goes best when you try to use it from the very beginning in authentic situations.

A.1 Getting Started

A.1.1 Why Python?

We have chosen to do our programming in Python because it is a popular programming language for mathematicians and scientists that is not part of an expensive software package. It also has extensive libraries of functions for mathematics and science. Python is not the only language with these properties, and if you already know another language, by all means use it.

A.1.2 Python versions 2 and 3

Python 3.5.2 is the newest stable version of Python as of this writing. It is important to note that Python 3 is not completely backwards compatible with Python 2. All examples presented in the text have been made to work with either version of Python. As a consequence, certain shortcuts specific to either version 2 or 3 have been avoided. For example, in Python 2, "print" can be called with or without parentheses, but in Python 3, parentheses are required. Since "print" with parentheses works in either version, we will use the parentheses.

A.2 Installation and Requirements

Python has no particular minimum requirements and will run on any major operating system (Windows, OSX, Linux).

A.2.1 Integrated Development Environments (IDEs)

Although Python scripts can be written in nearly any text editor, an integrated development environment (IDE) provides many advantages. One of the simplest (but also most useful) features is *syntax highlighting*. Syntax highlighting automatically changes the color of the function names, variables, keywords, punctuation, and literals that you type, giving you immediate visual feedback to syntax problems in your code.

We recommend Spyder as an IDE because it is free, there are versions for Windows, OSX, and Linux, and it has everything we need to get started. The default installation of Spyder also includes many Python libraries commonly used in math and science.

Installing Spyder on your computer is easy. Simply do an internet search for "installing the Spyder IDE." At the time of this writing, such a search led quickly to:

https://pythonhosted.org/spyder/installation.html,

which describes how to get Spyder and Python on a computer with just about any operating system. Simply follow the directions that apply to you. I put WinPython on my Windows machine and it has worked well.

A.3 Python Basics

A.3.1 Exploring in the Python console

Let's begin with integers, floats (numbers written as decimals, e.g., 2.0 and 0.0012), and arithmetic operations. Even before you write your first Python program, you can start to learn about variable types and operations by typing directly into the Python console. By default, Spyder has a Python console in the lower right portion of its window. If it is not visible, a new one can be opened through Spyder's menu Interpreters -¿ Open a Python interpreter. The Python command line begins with three greater than symbols where you can enter Python commands. Start with some integer arithmetic. Try out some addition, subtraction, multiplication, and exponentiation. Note that to compute a^b you must type $a**b$.

```
>>> 2+3
5
>>> 2-3
-1
>>> 2*3
6
>>> 2**3
8
```

Python also has a *modulus* operator, %, which finds the remainder when the first integer is divided by the second. For example,

```
>>>10%2
0
>>>10%3
1
```

because 2 divides evenly into 10 but 3 leaves a remainder of 1 when divided into 10. This operator may be new to you; we will discuss it in detail in Chapter 6 when we talk about modular arithmetic.

Let's look at integer division because not all consoles will produce the same output. Try 10 divided by 2 and 2 divided by 10 and see what happens. When I type into my console I get:

```
>>> 10 / 2
5.0
>>> 2 / 10
0.2
```

However, if I write a line of code in my editor that prints 2/10 to the console, I get 0. Depending on which packages are imported when you open your Python console, you might get 0.2 when you enter 2 divided by 10, or you might get 0. The latter happens because, unless some package overrides it, Python will return a value that is of the same type as those on which it operated. Because 2 and 10 are integers, Python will do the operation in the integers and return the integer part of the answer, which is 0 in this case.

To avoid this situation, let Python know you want to do the operation in the real numbers by making the values clearly floating point values instead of integers. Thus you should type

```
>>> 10.0 / 2.0
5.0
```

```
>>> 2.0 / 10.0
0.2
```

Because programming languages will take what you type literally, whatever output you obtained in our little experiment above, get in the habit of typing 2.0 instead of 2 whenever you are thinking about doing your arithmetic in the rational or real number system.

A.3.2 Your first programs

Usually you will not type directly into the console. Instead, you will want to type in the editor and then run the program, with the output displayed in the Python console.

Example A.1. The canonical first program in any language is the one that simply prints the string *Hello, world!* as output. Because you want to print this string and not the value of some variable that has been given the name *Hello, world!*, you must put quotes around the string. Thus, you should type the following in your editor:

```
print("Hello, world!")
```

Then run the code and observe the output in the console.

Exercise A.1. What appears in the console if you run the code without the quotes?

Of course, usually when you write a program, you are assigning values to variables, performing one or more operations, and outputting the result. Consider the following (still very boring) example.

Example A.2. In Python, the symbol = performs the assignment of a value to a variable. Assignment commands are evaluated from right to left. In this program, we assign the value 11 to the variable x and we assign the value 17 to the variable y. Then we take the integer assigned to x and the integer assigned to y, perform the indicated operation, and print the result. When we run the program, the integer 28 appears in the console.

```
x = 11
y = 17
print(x+y)
```

Example A.2 is kind of silly because we don't usually write programs to do one specific calculation. Instead, we write routines that can be applied to many different values of the variables. In such cases, we may want to prompt the user to enter the values on which to operate. Gathering such input from a user is easy.

Example A.3. When I taught high school, I was often asked, "If I have a 95% currently, what do I need to get on the final to get an A?" Let's write a program to answer this sort of question. We will ask the user to enter his current class average, the average with which he hopes to end, and the weight given to the final. We will report the necessary score he must earn on the final. The point of this example is to illustrate the syntax to use to get input from the user and assign it to a variable. In this program, we have three such variables.

```
# prompt the user for current average, desired average, and
    weight given to the final
current=input("Enter your current average as a number between 0
    and 100: ")
desired=input("Enter the average with which you hope to end as a
    number between 0 and 100: ")
weight=input("Enter the weight given to the final exam as a
    number between 0 and 100: ")

a=weight/100.0 #weight converted to a decimal
b=1-a #decimal representing portion of the grade already
    determined

needed=(desired - b*current)/a #finds score needed on the final

print(needed)
```

Exercise A.2. Write a tip-calculator program that prompts the user for a restaurant bill total and gives as outputs the values of a 15%, an 18%, and a 20% tip.

A.3.3 Good programming practice

Naming variables. Python allows you to create new variables at any time. Although you have a lot of flexibility when it comes to naming your variables, you should follow some basic rules:

1. Variable names are case-sensitive. The variables X and x are two distinct variables with no relationship to one another. Although sometimes in mathematics we may have two variables that differ only in case (e.g., we may let $T(t)$ be the temperature of a liquid at time t), we will try to avoid such variable names in our Python code. Errors caused by incorrect capitalization are very common and, if their effect is subtle, they can be very time-consuming to fix.

2. Variable names may contain only letters, numbers, and the underscore "_" character. They cannot start with a number.

3. There are keywords in Python that are reserved and cannot be used as variable names. These are: and, assert, break, class, continue, def, del,

elif, else, except, exec, finally, for, from, global, if, import, in, is, lambda, not, or, pass, print, raise, return, try, while, and yield. If you're using an IDE, these keywords will have a different color than your other variable names, which will tell you that Python will interpret them differently.

Comments in code. Observe that we have included comments in our code to describe what each piece does. Comments in Python must be preceded by the symbol #. This symbol indicates that what follows is to be ignored when the program is run. We recommend using comments liberally in your programs. They will help you to remember later what you were thinking when you wrote your code. They will also help others who are reading your code. Looking at code another has written without comments is like trying to understand a solution to a hard story problem just by looking at someone's scratch work. Communicate with your reader. We give this same advice when it comes to proof-writing. It is not a sign of intelligence to be able to write a proof or a program that no one can follow. But it does take real intelligence to be able to write clean, clear, well-organized code and proofs to solve challenging problems.

There is another important use for the symbol #. All too often, you will try to run your code and generate an error message. Sometimes it is easy to find and fix the error, but sometimes it is not. One approach to debugging is to temporarily "comment out" lines of code by putting # in front of them to see if running the code without those lines still generates an error message. If it does not, the error is in the part you commented out.

Operators and precedence: Parentheses are your friends. The mathematical operators have the usual precedence in Python; exponentiation happens before multiplication and division, which happen before addition and subtraction. But when writing code, because we enter mathematical expressions in plain text, parentheses are incredibly important. *It never hurts to use more parentheses.*

Although assignment happens from right to left, when operations have the same precedence, they are evaluated from left to right, e.g.,

```
>>> 3 / 2 * -5
-7.5
>>> 3 / (2 * -5)
-0.3
```

Division symbols are only one character, so very frequently the numerators and denominators need to be wrapped with parentheses. For example, the equation

$$x = \frac{-b + \sqrt{b^2 - 4ac}}{2a}$$

might be translated into Python as:

```
x = (-b + sqrt(b**2 - 4 * a * c)) / (2 * a)
```

When writing mathematics, it is acceptable to denote multiplication by juxtaposition, as in $4ac$. In Python, it is essential to indicate the operation, typing 4*a*c.

A.3.4 Lists and strings

With these basics in mind, let's talk about lists and strings. A *string* is just a collection of characters–letters, numbers, symbols, and punctuation. In our first program,*Hello, world!* was a string of characters we wished to print. The words were not names of variables. When we want to represent a string, we put it in quotes, as we did above. If we put in quotes some collection of characters that happen to appear elsewhere in the program, it doesn't matter–we still just have a string of characters. For example, if we add the line

```
print("x+y")
```

to the program from Example A.2, we get x+y as output even though earlier in the program there is a variable named x and it was assigned the value 11, etc.

We can perform the operation of concatenation on strings by using +. Thus the print command

```
print("x" + "y")
```

takes the string "x" and the string "y" and concatenates them. The output is xy.

Return to Example A.2. Suppose we want the program to have output

x + y is 28.

We must be careful, because part is a string and part is an integer arising from doing a calculation. We must tell our program to think of them both as strings and concatenate them. The following gives the desired output, as you should check.

```
print("x + y is " + str(x + y))
```

A *list*, on the other hand, is a variable type. The elements of a list can be integers, floats, or other lists. We put lists in square brackets, with the entries separated by commas. Here are two examples:

```
S = [1,3,5,7,9]
T = [
  [1,2,3],
  [4,5,6],
  [7,8,9]
]
```

The first list, S, has 5 elements, each of which is an integer. The second list, T, has 3 elements, each of which is itself a list with 3 elements. $S[i]$ denotes the

i-th element of the list, *but note that lists in Python are zero-indexed.* Thus for list S above,

$$S[0] = 1, \quad S[1] = 3, \quad S[2] = 5, \quad S[3] = 7, \quad S[4] = 9,$$

whereas

$$T[0] = [1, 2, 3], \quad T[1] = [4, 5, 6], \quad T[2] = [7, 8, 9].$$

We can perform operations on elements of lists. With S and T as defined above, consider the following operations:

```
x = S[0] + S[1] + S[2] + S[3] + S[4]
y = T[0] + T[1] + T[2]
```

Although the syntax looks similar, if we print x and y, we see that the results are very different. In the first case, each $S[i]$ is an integer, so our code asks that the five integers be added, with the result assigned to the variable x. If we print x, we see 25. In the second case, each $T[i]$ is itself a list, and the operation $+$ performed on lists concatenates them, forming a new longer list. Thus if we print y, we see $[1, 2, 3, 4, 5, 6, 7, 8, 9]$. If we instead want to think of T as a 3×3 matrix with first row $[1, 2, 3]$, etc., and we want to add all 9 entries, we use the notation $T[i][j]$ to refer to the j-th element of the i-th list and use the following command:

```
z = T[0][0] + T[0][1] + T[0][2] + T[1][0] + T[1][1] + T[1][2] +
    T[2][0] + T[2][1] + T[2][2]
```

In this book, we study sequences extensively and thus we generate many lists. We can specify a list by giving a formula for the i-th term or by giving a recursive definition, in which we specify the first term and then tell how to produce the i-th term from the previous. Both approaches require us to understand loops. Loops will be the topic of Subsection A.3.6.

A.3.5 *if . . . else* structures and comparison operators

Very often, we want what a program does next to depend on the result of some previous step. We use the *if. . . else* structure in these situations. Before we can describe this structure, we need to talk about comparison operators and blocks of code.

As we have already discussed, in Python, the equals sign $=$ is an assignment operator. Thus $a = 10$ in a program takes the integer 10 and attaches it to the label a. If we wanted to express this idea in a mathematical proof, we would probably say "Let $a = 10$." In mathematics, we are somewhat more used to thinking of the equals sign as it is used in a statement with a truth value. In mathematics, $10 = 20/2$ makes perfect sense; the integer 10 is the same as the integer obtained by taking 20 and dividing it by 2. In Python, however, the above makes no sense; we would be asking Python to take the result of the calculation 20/2 and assign it to a variable named 10. Because 10 is not a valid variable name, we would get an error message.

We might, however, want to test whether two calculations give the same result. In this case, we use the comparison operator ==. The result of this operation is either *True* or *False*, depending on whether the two values are or are not equal. Let's look at two comparisons performed in the Python console.

```
>>> 5 * 4 == 40 / 2
True
>>> 5 * 4 == 20 / 2
False
```

Comparison operators like ==, <, <=, etc., often appear in programs in combination with *if. . . else* structures. We illustrate with a simple program that takes two numbers, multiplies them together, and prints different messages depending on whether or not the product is less than 7.

```
a = 3.141592
b = 2.71812

if a * b < 7:
    print("The product is less than 7.")
else:
    print("The product is greater than or equal to 7.")

print("The result is " + str(a * b))
```

Observe that, in Python, we use a colon at the end of the *if* line and the *else* line and we indent the blocks of code that are to be executed in each case. We have also added a line of code that is *not* indented that will print the actual result of the calculation. If we had indented this line, it would only be executed if $a * b < 7$ is false.

Exercise A.3. Write a program that prompts the user to enter two positive integers and then determines whether their product is divisible by 12.

We can also combine multiple comparisons with the keywords *and* and *or* to do more complex logic.

Example A.4. When you prompt a user for input, it is often a good idea to use one or more conditional statements to check that their input is valid. For example, suppose your program prompts the user for two positive integers a and b with $a < b$. The test of validity of the input might look like this:

```
if a<= 0 or b<=0 or a>=b:
    print("Invalid input.")
```

We can also have three or more branches for the program to take by using an *if. . . elif. . . else* statement. *elif* is short for *else if*.

Example A.5. We illustrate by writing a program that looks for the smallest prime factor of an integer. For simplicity, we will only check for divisibility by the first few primes.

```
x=input("Enter an integer: ")

if x%2==0:
    print("2 is the smallest prime factor of x.")
elif x%3==0:
     print("3 is the smallest prime factor of x.")
elif x%5==0 :
     print("5 is the smallest prime factor of x.")
else:
    print("x is not divisible by 2, 3, or 5. ")
```

We use such a structure when we want to execute *precisely one* of many blocks of code. As soon as one of the conditionals is true, the corresponding block of code is executed and no other conditionals are considered. To illustrate this point plainly, let us consider what the program does when the input is 15. The conditional x%2==0 is false, and so we proceed to the first *elif* line. The statement x%3==0 is true, and so the corresponding block of code is executed, printing the statement "The smallest prime factor of x is 3." Because we found a true conditional, we do *not* proceed to the next *elif* statement. Thus even though x%5==0 is also true, this block of code is not executed.

Exercise A.4. Modify the program in Example A.5 so that the program reports *all* of the primes less than 10 that are factors of a given integer.

Exercise A.5. Write a program that takes as input the coefficients of a quadratic polynomial in standard form and reports the number of real roots it has.

A.3.6 Loop structures

The *for* loop. We use a loop whenever we need to perform a set of instructions repeatedly, for example when we are generating the first terms of a sequence. Because we spend a lot of time in this book studying sequences, it is very important for you to master loops. We give lots of examples and exercises.

Example A.6. As our first example, we generate the first 15 elements in the list of perfect squares. Let's name our list S. In this case, we have an explicit formula for the i-th term, namely

$$S[i] = (i+1)^2.$$

(Remember that lists are zero-indexed!)

To generate S, we begin with an empty list. We then go through a loop 15 times. Here is the syntax:

```
for i in range(0,15):
```

(Note the colon at the end of the line.) This line indicates that we have some procedure we want to execute a number of times, once for each index i in the indicated range. The starting index is $i = 0$. The index i will be incremented by one after each time the indented block of code (as yet to be shown) is executed. The process will continue as long as i is still *strictly less than* 15. Thus the block of code after our *for* statement will be executed for $i = 0, 1, 2, \ldots, 14$.

At each step i, we wish to add the i-th term of the sequence to our list S. We use the list function *append*. We call our list's append function using "dot" notation. Here's the entire program. Note that the block of code to be executed each time through the loop must be indented.

```
S=[] #start by defining an empty list

# generate terms for i = 0, 1, 2,..., 14
for i in range(0,15):
    S.append( (i + 1)**2 )

print(S)
```

Exercise A.6. (a) Generate the first 20 terms in the sequence given by $S[i] = 1.0/(i+1)$.

(b) What happens if you use $1/(1+i)$ in your program instead of $1.0/(i+1)$? Explain!

Next, let's look at an example that takes a non-negative integer a and computes $a!$. Recall that $0!$ is defined to be 1, and that, for $a \geq 1$, $a!$ is the natural number obtained by multiplying together all the natural numbers less than or equal to a. Thus $1! = 1$, $2! = 1 \cdot 2 = 2$, $3! = 1 \cdot 2 \cdot 3 = 6$, and so on. Let's write a program to find $8!$. We write it in such a way that we could use it to compute factorials of other natural numbers by just changing the value of a in the first line of code. Thus we begin by setting a equal to 8. We define a variable called *result* that is initially set to 1 because 1 is the multiplicative identity. Next, we go through a loop 8 (or a) times. The line to be executed each time is simple; we just replace *result* with the previous result times the next natural number $i + 1$. We end with a print command. Here's the entire program.

```
a = 8 #the number whose factorial we want to find
result = 1 #initialize result to 1, the multiplicative identity

for i in range(0,a):
    result = result*(i+1)

print(result)
```

Exercise A.7. The i-th triangular number, $T[i]$, is the sum of all natural numbers less than or equal to i. Thus $T[0] = 0$ (there are no natural numbers less than or equal to 0), $T[1] = 1$, $T[2] = 1 + 2 = 3$, $T[3] = 1 + 2 + 3 = 6$, etc. Write code to generate $T[n]$ for $n = 30$.

The *while* loop. Loops can be constructed in other ways. For example, sometimes rather than running through a loop a fixed number of times, we want to run through it as long as some certain condition is met. The *while* loop is good for this situation.

As an illustration, let's write a second program to compute $a!$, this time using a *while* loop. We start out the same way, setting a equal to 8 and *result* equal to 1. Each time through the loop, we take *result* and replace it with the old result times a natural number. The first time through, we multiply by $a = 8$. The second time though, we want to multiply by 7. We accomplish this by replacing the old a with $a - 1$. We keep going through the loop until the updated a is no longer positive, and then we print the result.

```
a = 8 #the number whose factorial we want to find
result = 1 #initialize result to 1, the multiplicative identity


while a>0:
    result=result*a
    a=a-1

print(result)
```

Exercise A.8. What happens if you replace the last line of this program with

```
print(str(a)+"! is: "+str(result) )?
```

Explain.

Exercise A.9. Using a *while* loop, write a program that finds the running total when the natural numbers $1, 2, 3, \ldots$ are added and terminates when the sum first exceeds 1000. Have the program print the sum at this point.

A.4 Functions

Because we often want to do the same calculations many times with different inputs, it is common to write *functions* that can be called repeatedly throughout a larger program. Just like the functions you meet in mathematics, functions you write in Python will take one or more inputs and return some unique output. We will illustrate with a number of simple examples. In some of these cases, Python already has a built-in function that does what we want. We will write our own function anyway since our aim is to use simple examples to illustrate the basic idea and to demonstrate proper syntax.

Example A.7. Let's start by writing a factorial function. In the first line, we name the function we want to define and indicate how many arguments it will have. In this case, we have decided to name our function *factorial*, and our function has one argument, which we have named *number*. Note the colon after the first line. The indented block of code says how to compute the factorial of *number*. We have seen this code previously. The last line of the function states that the output of the function is to be *result*.

```
def factorial(number):
    result=1
    for i in range(0,number):
        result=result*(i+1)
    return(result)
```

If you run this code, nothing will happen. We have defined a function, but we have not asked that it be evaluated for any particular input. If you now want to use it to see the factorial of a number you input, you could add two lines of code:

```
n=input("Enter a non-negative integer: ")
print(factorial(n))
```

Note that, although we used the variable name *number* in the definition of the factorial function, when we call the function, the input may have a different name. The variable *number* is a *local variable*; it is used within the definition of the function only. If somewhere else in the program we have a different variable with the name *number*, the value assigned to that variable will not affect what happens when the function is called. This phenomenon of local variables may initially seem odd, but it is really nothing new to you; when you define functions in calculus like $f(x) = x^2$, you know that x is just a place-holder and that the function under consideration is "the squaring function." When you want to use it, you can give the input a different name like t or y. Furthermore, if somewhere else on your paper you have used the letter x in a different equation, you don't worry that the other equation has affected your definition of f. Still, it is probably good practice when writing code not to use the same names for variables in function definitions that you use elsewhere in your program as you may confuse yourself, especially if the program is long.

The advantage of writing functions is that then you can call them repeatedly in some other larger program.

Example A.8. For non-negative integers n and k with $k \leq n$ we define a number $C(n, k)$ by the formula

$$C(n, k) = \frac{n!}{k!(n-k)!}.$$

This number is read "n choose k" (for reasons we will discuss in Chapter 5) and is called a binomial coefficient. We agree that $C(n, k) = 0$ if $n < k$. For fixed n, the $n + 1$ numbers $C(n, k)$, $0 \leq k \leq n$ form the n-th row of a triangular array known as *Pascal's Triangle.* We use the factorial function from Example A.7 in a program that defines a binomial coefficient function and then prints a row of Pascal's Triangle.

```
# finds the factorial of a non-negative integer
def factorial(number):
    result=1
    for i in range(0,number):
        result=result*(i+1)
    return(result)

# finds the binomial coefficient for an ordered pair of
     non-negative integers
def binom(first,second):
    if first<second:
        return(0)
    else:
        return(factorial(first)/(factorial(second)*factorial(first
             - second)))

# generates a row of Pascal's Triangle
def Pascal(row):
    S=[] #define an empty list to hold the entries of the row
    for i in range(0,row+1):
        S.append(binom(row,i))
    return(S)

# prompts the user and prints the desired output
n=input("What row of Pascal's Triangle would you like to see? ")
print(Pascal(n))
```

If you run this program and request to see row 4, you will see the output

$$[1, 4, 6, 4, 1].$$

Let's look at a simple function on a list.

Example A.9. We write a function that takes as input a list and a key and returns either -1 if *key* is not found on the list or the index of the first occurrence of *key* if it is found. In this program, we use a built-in function *len* that takes a list as input and returns its length (number of elements). We also use the *break* command to terminate the loop if we find *key* on the list.

```
def basic_search(list,key):
    index=-1 #initialize index to be -1
    for i in range(0,len(list)):
        if list[i]==key: #if key is found, update index and stop
            index=i
            break
    return(index)
```

Exercise A.10. Write a function, *reverse*, that takes as input a list S and returns the list in reverse order. For example, if $S = [1, 2, 3, 4]$, the output of *reverse* should be $[4, 3, 2, 1]$.

Exercise A.11. Write a function that takes as input a list and a key and counts the number of occurrences of the key on the list.

A.5 Recursion

Recall that when we define a sequence $\{a_n\}$ recursively, we give an initial condition and a rule that relates the value a_n to one or more previous values. Recursive algorithms are similar. They begin by specifying the output of the function for some simple input, and then they give a rule for relating the output for general input to output for simpler input. The best way to understand how recursion works is through examples.

Example A.10. We return to the factorial function, this time using recursion. The recurrence relation for factorials says that $0! = 1$ and, for $n \geq 1$,

$$n! = n \cdot (n-1)!.$$

The recursive algorithm also incorporates both these statements.

```
def factorial(number):
    if number==0:
        return(1)
    else:
        return(number*factorial(number-1))
```

Let's consider how this function finds 3!. Because $3 \neq 0$, we do not execute the *if* statement but rather proceed to the *else* statement. The function is to return $3 * \textit{factorial}(2)$. We have thus called the function again, with the smaller input of 2. Because $2 \neq 0$, we again do not execute the *if* statement but proceed directly to the *else* statement, where the function returns $2 * \textit{factorial}(1)$. Thus we call the function again with the input 1. Because $1 \neq 0$, we proceed to the *else* statement yet again, and the function returns $1 * \textit{factorial}(0)$. We call the function one last time with input 0. This time the conditional in the *if* statement is true and the function returns 1. Piecing together all these steps, we see

$$\begin{aligned}
\textit{factorial}(3) &= 3 * \textit{factorial}(2) \\
&= 3 * 2 * \textit{factorial}(1) \\
&= 3 * 2 * 1 * \textit{factorial}(0) \\
&= 3 * 2 * 1 * 1 = 6.
\end{aligned}$$

Exercise A.12. The Fibonacci sequence $\{F_n\}$ is defined by the recurrence relation $F_0 = 0$, $F_1 = 1$, and for $n \geq 2$, $F_n = F_{n-1} + F_{n-2}$. Write a recursive algorithm that takes as input a non-negative integer n and returns F_n.

Index

A

B

C

D

E

F

N

O

P

T

U

V

W

Z